A Leaner's Guide to Fuzzy Logic Systems

A Leaner's Guide to Fuzzy Logic Systems

Dr. K. Sundareswaran

CRC Press is an imprint of the
Taylor & Francis Group, an **informa** business

CRC Press
Taylor & Francis Group
52 Vanderbilt Avenue
New York, NY 10017

CRC Press is an imprint of Taylor & Francis Group, an Informa business

Printed on acid-free paper

International Standard Book Number-13: 978-0-367-25443-8 (Hardback)

Library of Congress Cataloging-in-Publication Data

LoC Data here

Visit the Taylor & Francis Web site at
http://www.taylorandfrancis.com

and the CRC Press Web site at
http://www.crcpress.com

This book is submitted to the
lotus feet of Lord Krishna

Contents

List of Figures

List of Tables

Preface to Second Edition

SINCE PROFESSOR ZADEH'S FIRST paper on fuzzy sets in 1965, the field of fuzzy logic systems has become one of the most interesting and fastest-growing technologies in the world. The concept of fuzzy logic has emerged as an alternative method to conventional theory in dealing with systems where uncertainty prevails. Such systems include engineering applications, economics, business, bio-medical applications, etc.

A Learner's Guide to Fuzzy Systems is primarily intended for undergraduate students and researchers to facilitate education in the ever-increasing field of fuzzy logic. The book begins with uncertainty-related issues in Chapter 1, and the concept of fuzzy sets and their operations are systematically illustrated in Chapter 2. The methodology of rule base design together with various fuzzy reasoning schemes is demonstrated in Chapter 3. A chapter on fuzzy logic design is incorporated for researchers pursuing higher studies in this field.

This book is suitable for self-study as well as a textbook for regular coursework.

Author

Dr. K. Sundareswaran earned his MTech (Hons) in power electronics from the University of Calicut, Kerala, India, and a PhD from Bharathidasan University, Tiruchirappalli, Tamil Nadu, India. From 2005 to 2006, he was a professor with the Department of Electrical Engineering, National Institute of Technology Calicut. Dr. Sundareswaran currently works as a professor (HAG Scale) in the Department of Electrical and Electronics Engineering, National Institute of Technology, Tiruchirappalli. His research interests include power electronics, renewable energy systems and biologically inspired optimisation techniques.

CHAPTER 1

Unravelling Uncertainty Through Simple Examples

1.1 INTRODUCTION

This chapter is a sincere effort to explore different perceptions of the term "uncertainty". The concept is explained so as to be understood by an engineering professional as well as a layperson. Simple examples are employed as a platform for explaining the term.

1.2 EXAMPLES

The Oxford Dictionary defines the meaning of the word uncertainty as "not knowing definitely". Other descriptions include imprecision, inaccuracy, vagueness, inconsistency, ambiguity, unpredictability, etc. A layperson, however, is not worried about the exact meaning of uncertainty and the terms "uncertainty", "ambiguity", "vagueness", etc. are largely used in everyday conversations rather than in the technical field. Such words and phrases are used interchangeably in different contexts without much thought. Before going into the significance of uncertainty, consider the following examples.

Example 1

Consider the issue of speed control of a separately excited direct current (dc) motor employing an armature voltage control, as shown in Figure 1.1(a). The objective is to achieve a desired motor speed under different load conditions. The variation in motor speed with different values of armature

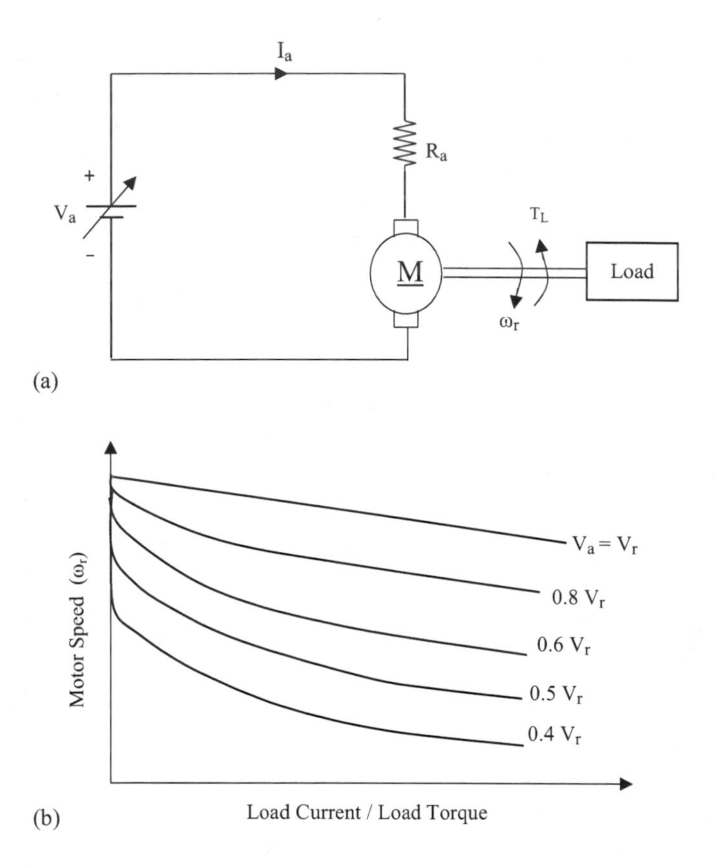

FIGURE 1.1 Direct current motor speed control: (a) circuit and (b) torque-speed characteristics.

voltages against a load torque for a typical dc motor is plotted in Figure 1.1(b), neglecting the armature resistance variation and aging of the motor field.

Figure 1.1(b) shows that each speed-torque curve is almost a straight line with a negative slope and all such lines are parallel to each other. Here, V_r stands for rated armature voltage. This perfectly represents a linear system. For speed controller design, one needs to have a "model" of the dc motor, and in most cases, the model is a transfer function or a state-space representation. This model is time invariant and linear. The closed loop controller design can be realised using the linear control theory. The dynamic responses of the system at all operating points resemble each other. If the speed response is taken as the objective, then a good dynamic response is always achieved at all operating points of the motor.

Example 2

Consider the speed control of the same dc motor mentioned in Example 1; however, this time the motor armature is supplied from a silicon controlled rectifier (SCR) converter, as shown in Figure 1.2(a). Here, the armature voltage is controlled by varying the SCR firing angle, α. The variation in the motor speed against the load torque with several values of α is shown in Figure 1.2(b).

Clearly, unlike in the previous case, the characteristic curves are parallel only under rated torque conditions. Under light torque conditions, the curves are more steep and bent. This is attributed to discontinuous armature current under low torque conditions. Indeed, there are basically two modes of operation, namely, continuous and discontinuous modes with an SCR converter–fed drive. With continuous armature current, the drive system is described by two different sets of differential equations. Under the light

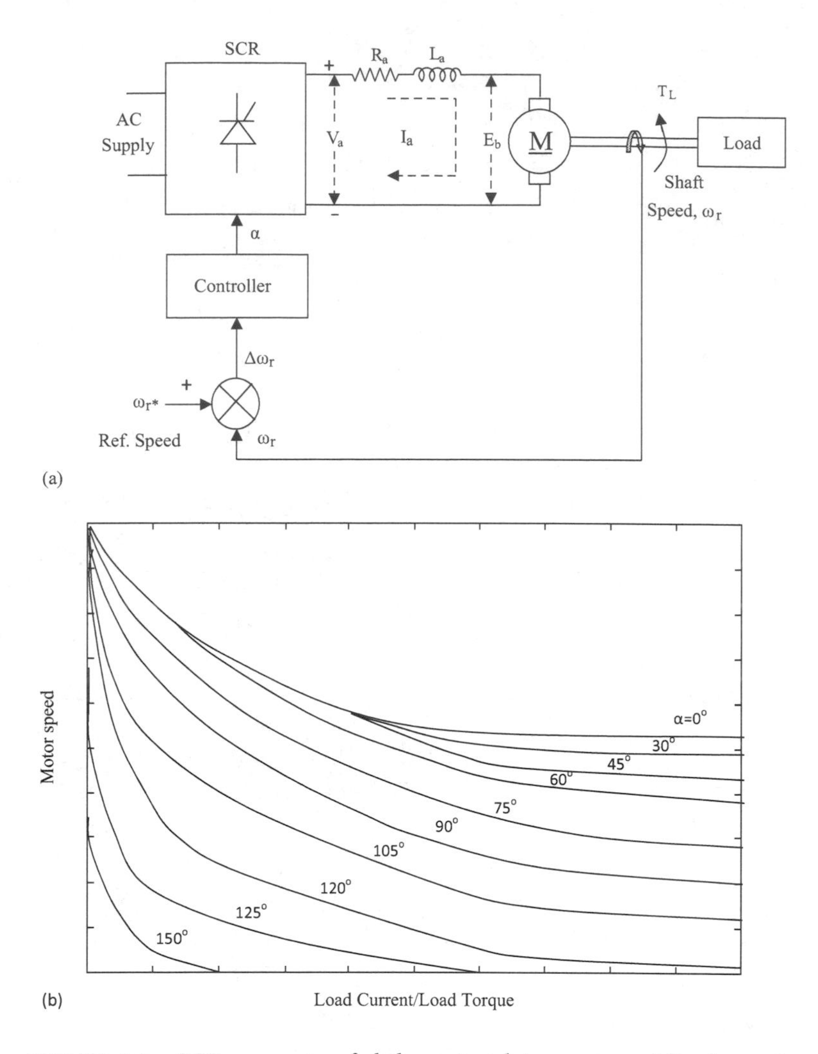

FIGURE 1.2 SCR converter–fed dc motor drive system: (a) schematic diagram and (b) torque-speed characteristics.

load condition, the armature current is discontinuous and torque-speed curves have different slopes. If a feedback controller is designed for a typical operating point, it will produce a good/satisfactory response at that point; at all other points, the fixed gain controller performance need not be good/satisfactory. Rather, the response will differ largely with a change in the torque-speed curve.

Example 3

Consider the process of washing clothes. Evidently, the process does not have a model, but we follow three steps sequentially: soak, wash and rinse. The time period for each stage depends on how dirty the clothes are and the type of dirt on them. The quantity of washing powder required, too, depends on the aforementioned two factors. On many occasions, we repeat the three stages of washing until the clothes are clean.

Example 4

Consider the case of a final-year undergraduate student applying for admission to a suitable postgraduate (PG) programme in his/her native country or abroad. The objectives to be considered while looking for a programme are

(a) Reputation of the institution,
(b) Availability of financial assistance,
(c) A master's programme of his/her interest and
(d) Employability of the programme.

To start the search, the student interacts with many seniors and faculty members of his/her current institute. The student explores online and communicates with potential professors abroad. At the end of the process, owing to sheer determination, the student secures admission. However, in most cases, it is uncertain whether all objectives are completely fulfilled. This is because, apart from the grades required for a postgraduate programme and entrance test scores, numerous factors such as the foreign policies of different countries, the availability of funds for financial assistance and the job market affect admission. Uncertainty also arises due to the multidimensional aspects of the objectives previously listed. Unlike the first two examples, no model exists for this case study. Most of the inputs are simple statements with no precision. Some of the statements could be partially true and a few others could be wrong. It depends on the ability/intelligence of the student to absorb the truth content and guide his/her search towards the best results.

In order to discuss the concept of uncertainty, let us initially define uncertainty as a measure of the guarantee of achieving a desired goal in a process. In the first two examples, our goal is to achieve a good dynamic response at all operating points. In the third example, the desired goal is to remove all dirt/stains from clothes. The goal in the fourth example is to gain admission to a reputed institute with financial assistance.

A closer observation of these four examples reveals that the first one represents a linear system. The dynamic responses of the motor speed at different operating points are identical, and no uncertainty is associated with them. In the second example, there are two/three state-space models in one complete cycle of operation leading to non-linearity. A fixed gain controller produces different transient responses at different operating points. One might think that "little uncertainty" is present in the dynamic responses under different load conditions. To make this aspect clearer, consider a motor drive that is used to move a passenger lift up and down. Let the full capacity of the lift be, say, eight passengers. With the set-up in Figure 1.1(a) in Example 1, irrespective of the number of passengers, a fixed gain controller produces a uniformly good dynamic response, i.e. the starting and stopping of the lift are always smooth and are independent of the number of people in the lift cabin. On the other hand, with an SCR converter–fed motor, a fixed gain controller always produces a good, satisfactory start-up/braking response in the case of four to eight passengers. However, oscillations or jerks are experienced during starting as well as braking of the lift cabin in the case of one to four passengers, as this is a light-load condition. Thus, in the second example, there is an uncertainty about the smooth functioning of the lift.

The third example has no defined model but has a practiced way of processing. Here, the objective is to make the clothes dirt free, and to achieve this goal, one needs to increase/decrease the number of stages of washing or repeat them a few times. However, there is no guarantee of the finishing time of the washing process or the level of cleanliness achieved. In other words, there is an increased level of uncertainty to achieve the goal of clean clothes. However, a clothes launderer faces very little uncertainty about the cleanliness of clothes since he has vast experience of washing different types of clothes with different varieties and amounts of dirt. When the experience of a clothes launderer is integrated with a commercially available washing machine, the machine becomes "automatic" and is more "successful" in reducing the uncertainty associated with washing clothes. An automatic washing machine is one of the best examples of a system making use of human experience.

The fourth example is more complex. Naturally, it has no model; further, the processing needs information from several individuals. Much of the information will be incomplete; some of it may be partially true or distorted. The student aspiring higher studies must judicially extract the essence of all the information and proceed. In this case, uncertainty is more compounded.

It is interesting to observe that most of us face challenges/tasks that are very similar to Example 4. The issues we face and ultimately sail through do not have any representation such as mathematical models or block diagrams. We receive inputs from our surroundings, process them with the help of experience and reach our goals. The quantum of uncertainty in each task largely depends on the multiple dimensions of our objectives, the quality of inputs and our ability to process the inputs.

1.3 A SIMPLE VIEW OF FUZZY LOGIC

Consider that you are asked to compute the value of x given by

$$x = 1.68 \times 10^{-3} + \frac{84.18 \times 10^{3}}{128.9} + \frac{17.23}{4.2 \times 10^{-2}} + \frac{78.83}{24.8 \times 10^{2}}$$

Definitely, the computation will demand at least a couple of minutes, even if you use a calculator. However, present-day computers will take only a fraction of a second to compute x.

Now consider another case, where you are asked to compute y given by

$$y = (\text{around } 100) + \text{less than } 10$$

There are numerous solutions such as 109, 108, or 107. All such solutions are valid and it takes only a few seconds to arrive at one. But a computer will never give a solution as it cannot understand "around" or the precise meaning of "less than".

This example illustrates that although computers process numerical data at a very high speed, they fail to understand imprecise data. On the other hand, human intelligence has a poor computing speed but can process imprecise data quickly and can reach a logical conclusion. Interestingly, human intelligence does not prefer precise data and always looks for approximations. Thus, if human intelligence is imparted to computers, it is possible to perform several operations successfully such as closed loop control of many processes that either do not possess a model or possess incomplete non-linear models. Fuzzy logic is one such tool that enables machine (i.e. computer) learning. Fuzzy sets (fuzzy membership functions) and fuzzy set operations are most suitable for the machine learning process.

1.4 LEARNING ABILITY

For realising machine learning, it is important to study human intelligence, its properties, adaptability, etc. It was this background that led to the birth of artificial intelligence, popularly known as AI. This is a discipline that aims to understand human intelligence and impart it to a computer program.

The beauty of human intelligence is that it tries to learn from experience. It can be stated that one possesses intelligence with 100% rigour if one can perform the following tasks:

(a) generalising, (b) drawing analogies and (c) selectively discarding irrelevant information.

(a) Generalising: A computer stores numerical data perfectly despite the randomness of the data. However, the human mind tries to look at the salient features of the data and memorises these features. For example, consider the following marks obtained by a group of students in an examination (out of 100):

 28, 89, 62, 77, 52, 91, 56, 72, 79, 87, 73, 64, 71, 77, 64, 69, 73, 64, 73, 72, 61, 75, 53, 79, 74, 68, 70.

 The major observations from these results would be: many have done well with marks in the range 65–80; only one student got a low mark of 28; one student with 91 marks did very well, etc. Thus, rather than storing information, the human mind extracts the most noticeable facts and preserves them.

(b) Drawing analogies: For making decisions, either major or minor, the human mind always recalls past, similar experiences/incidents/circumstances to arrive at the right decision. The past experiences could be that of an individual or those associated with the individual. While travelling, suppose a person is stranded for some reason and is unable to reach the destination; he/she recalls similar experiences of his/her own or of those associated with him/her and, considering the present state of affairs, tries to stitch together a reliable solution to reach the destination. Thus, a possible solution could be achieved by suitably amending a past experience to the present situation. This is commonly seen in the medical profession.

(c) Selectively discarding irrelevant information:

The essence of the learning process basically stems from this property – forgetfulness. Forgetfulness, considered as a weakness of the human mind, is indeed a blessing in disguise. The human mind possesses an immense capacity to learn from forgetfulness, otherwise treated as a bane. Consider the following simple questions and answers:

(i) What was the colour of your shirt on the first day of your college?

The answer from almost all people would be "I don't know". Very rarely, some may say, "I think the colour was _________".

(ii) How many boys and girls were there in your 12th class?

The answer could be "I think ___ girls and ____ boys".

(iii) What is your name?

An individual would precisely tell it! The uncertainty observed in the answers to the previous two questions doesn't apply to this one.

The human mind preserves only that information which is relevant. Unwanted or least significant information is wiped from the mind.

1.5 DIFFERENT PHASES OF UNCERTAINTY

While the term "uncertainty" has been used in the previous sections, terms such as "ambiguity", "vagueness", "imprecision", "inaccuracy", "inexactness", and "fuzziness" are also commonly employed in such circumstances. It is interesting to discuss these terms and classifications.

1.5.1 Inexactness

This term refers to our inability to measure variables in a precise manner. Generally, in scientific or engineering applications, the measurement of various parameters is common and such measurement is subject to a large degree of uncertainty due to various errors attributable to either instruments or humans. Instrument errors can occur for a variety of reasons such as calibration, non-uniform scale, aging, and change in room temperature or humidity. The main cause of human error in measurement is due to parallax error, which arises from a change in the measurement value with a change in the position of the observer.

In the field of measurement, precision and accuracy are used interchangeably, as the two are related concepts. However, there is a fundamental difference between the two. Precision refers to a number of digits/decimals representing the measurement and depends on the granularity of the scale, i.e. the finest division restricts the precision. Thus, a measured current of 3.1098 A is more precise than 3.1 A. On the other hand, accuracy measures the degree to which a measured value coincides with the "true" or "actual" value of a variable. Accuracy is the difference between a measured result and the true value. If the actual power consumed by an incandescent lamp is 100 W and if a wattmeter measures 101.8 W, the accuracy of the instrument is −1.8 W. Thus, the accuracy of an instrument is affected by external and internal influences, and precision is affected only by a number of divisions on the scale.

1.5.2 Semantic Ambiguity

Ambiguity generally refers to the property of possessing several distinct but equally plausible and reasonable interpretations of a particular state of an event. By semantic ambiguity, we mean the ambiguity associated with the meaning of a word, phrase or sentence.

Here are two examples:

1. You are driving a car with your friend in the passenger seat. You reach a junction, but are unaware of the direction of your destination. You ask: "Shall I turn left?" Your friend answers, "right". Here, the answer may simply mean "you are right" (so turn left) or "no, you have to turn right". The interpretation depends on the tone and tenor of your friend's voice, body language, etc.

2. You go with your family to eat at a restaurant. Your wife says, "The food is hot". The term "hot" may mean that the temperature of the food is high or that it contains too much chilli and is thus hot.

1.5.3 Visual Ambiguity

This ambiguity arises over the position, location or trajectory of an object or a system due to the representation of the object, the observer's position with reference to the object or the relative velocity between the two. Consider point P marked in Figure 1.3. The point can either be on one of the sides ABDC or EFHG or within the volume. The uncertainty about the point is due to our inability to represent the object in the 3-D plane.

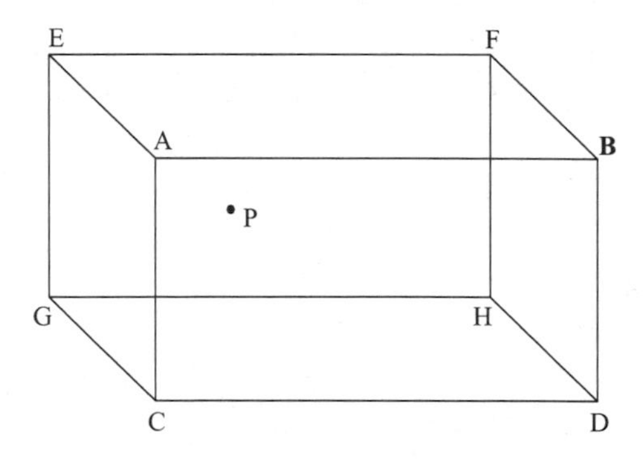

FIGURE 1.3 Example of visual ambiguity.

Consider a rotating shaft observed by two people standing on either side of the shaft. For the first person, it rotates in a clockwise direction, while for the second person, it rotates in an anticlockwise direction. This difference is because the location of the observer changes with reference to the rotating object. When you travel on a bus, you feel that trees and buildings are moving past you. However, the fact is otherwise. Visual ambiguity hinders the ergonomic design of systems and components.

1.5.4 Structural Ambiguity

The interconnections and interactions of different components of a system can cause a high level of vagueness. This can be attributed to an increase in the number of components, similar names, lack of definition of each component, etc. Consider the case of a family tree where names such as father, mother, son, and daughter appear multiple times. As the same name and function are repeated, the uncertainty of the system configuration looms large. Another example is a power system network linking many generators, transformers (transmission, distribution, etc.), transmission lines, impedance values, different loads, etc. In the case of a lengthy computer program, structural ambiguity may arise when there are several subroutines with the same/similar names and functions.

1.5.5 Undecidability

A deeper form of ambiguity originating from our inability to discriminate between different states of an event is termed "undecidability". If an event remains unchanged, there are different possible and reasonable states of the event, causing uncertainty for any one state. This uncertainty is called undecidability. In an undecidable state, we are unable to make a reasoned choice between ambiguous representations of a model. Here are a few examples.

A final-year engineering student is offered two jobs through campus placement. One is in a hardware engineering firm, whereas the second job is in a software development organisation.

The student is in a dilemma due to the various pros and cons of both jobs. The job at the hardware engineering company is stable. Additionally, development in the field is slow, and he can update himself at a comfortable pace. He can attain expertise in that field and reach higher posts in the company, or he can switch to similar companies. However, the salary is lower than that in the software firm, and the work environment is not attractive or conducive to physical comfort. Physical work may be involved, and there are few companies in that area, thereby limiting his choice of switching to other organisations. The software job, on the other hand, has an attractive salary and an air-conditioned workplace. Additionally, there is less physical work and the presence of a large number of companies in the area will give him the option of switching to other companies. The drawbacks are that he has to update his knowledge very frequently, and the job stability depends on global/market requirements, making him compete with people from non-engineering streams. Thus, to fine-tune his decision, he needs to consider a large number of variables.

Another example is the choice of motor for a variable-speed application. Generally, there are two choices: a dc motor and an induction motor. With a variable armature voltage and/or a field current, a dc motor can operate at any torque-speed point; however, dc motors are costlier, bulkier and incur greater maintenance costs. On the other hand, induction motors possess constant speed characteristics, and to get a variable speed, inverter circuits with complex control algorithms are required. Moreover, induction motors are cheaper, more rugged and incur practically zero maintenance costs. Thus, the mind of an electrical engineering professional oscillates between the two choices. In a similar manner, a half-open door is equivalent to a half-closed door. At first sight, the door appears half-closed; then it reappears half-opened. In such cases, our mind toggles from one state of the event to the next and so on, without sticking to any state permanently.

1.6 PROBABILITY AND UNCERTAINTY

Some theoreticians still believe that probability theory can well unravel an uncertain situation. However, the counterargument is that probability theory is an attempt to explain how events occur in a random space. The basic tenet of probability itself is the randomness of events. Thus, the representation requires knowledge of this random space as a prerequisite. For example, when a coin is tossed, we know that the probability of getting heads is half. Here, since we know the total solution space, we are able to predict the output probability. Further, the solution is a subset of this random solution space. In many cases, the solution space may not be available as a subject. In the case of predicting climate, there are a large number of subsets, but these are not fixed subsets. These subsets vary in dimension, complexity and characteristics. This leads to an output, which is a "combination" of many subsets.

Second, probability is concerned with a population; it does not consider individual instances. Consider that 10 final-year students are shortlisted for interview by a reputed company. The company only requires one candidate from the college. Accordingly, the probability of each student is 1/10. But all students may not be equally good in all aspects. The group of 10 students comprises bright as well as average students. Say, one student has undergone an internship in the field in which the company is recruiting people. He may have a very bright academic background and good communication skills. This student has the maximum probability of getting the job, and in practice, his probability is more than a mere 1/10. Thus, when probability is applied to an individual instance, it simply vanishes. Probability tells us something about a population but not about individual instances.

Third, probability is a solution to uncertainty associated with time. In the previous example, until the interview results are declared, the solution to uncertainty is 1/10. Once the result is announced, there is no uncertainty, and the solution, 1/10, has no meaning associated with it. This leads to the conclusion that the solution offered by probability is time-bound.

1.7 CONCLUSION

The concept of uncertainty and related topics have been discussed in this chapter. It has been shown that all events have a certain degree of uncertainty associated with them. Further, the concept of probability fails to decipher the level of uncertainty associated with events.

QUESTIONS

1. Describe the concept of uncertainty with appropriate examples.
2. Name a process in which the extraction of an exact model is difficult. Also, specify the parameters of the plant to be identified.
3. Describe major difficulties associated with the controller/estimator design of an uncertain system by using a suitable example.
4. Illustrate the concept of uncertainty in a non-linear system.
5. Write a brief note on vagueness associated with man-machine systems.
6. Give an example of a system where transfer function derivation is strenuous.
7. What is the basic difference between a conventional system design and an artificial intelligence–based system design?
8. What is meant by model-free systems?
9. Write a short note on expert systems.
10. Write a brief note on the properties of human intelligence and artificial intelligence programmes.
11. What is meant by "partial" information? Illustrate.

12. Elaborate on the decision-making ability of humans with partial/untrue information.
13. What is meant by knowledge-based systems?
14. Write a brief note on artificial intelligence.
15. Outline the major features of human intelligence.
16. Write about the major disadvantages of the expert system technique.
17. Define ambiguity in the context of fuzzy systems.
18. Define inexactness.
19. Explain precision and accuracy in instruments.
20. What is inaccuracy? How is it caused?
21. How will you increase the precision of an instrument?
22. How will you make a voltmeter precise?
23. Using suitable examples, illustrate semantic ambiguity.
24. Illustrate visual ambiguity and mention a few application areas.
25. Define and illustrate structural ambiguity with necessary examples.
26. Examine the structural ambiguity associated with a three-phase thyristorised ac/dc converter.
27. Investigate the quantum of structural ambiguity associated with a transistorised astable multivibrator circuit.
28. Discuss structural ambiguity associated with a dual converter–fed dc motor drive.
29. Analyse the structural ambiguity associated with static Scherbius drives.

30. Define undecidability and illustrate with necessary examples.
31. Analyse the undecidability of an individual who is settling either in his/her native town or a metropolitan city.
32. Explain how probability fails to unravel uncertainty.
33. Can statistical means solve uncertainty? Explain.

CHAPTER 2

Fuzzy Sets

2.1 INTRODUCTION

This chapter explains various definitions and concepts associated with fuzzy sets. To distinguish between fuzzy sets and conventional sets (non-fuzzy sets), the latter is referred to as crisp sets in the associated literature. Crisp sets are introduced first and their limitations are highlighted. Different types of commonly employed fuzzy sets and various operations using fuzzy sets are illustrated.

2.2 CLASSICAL SETS (CRISP SETS)

The classical set theory was founded by the German mathematician George Cantor. A classical set, which is called a crisp set in the fuzzy logic context, is a collection of objects of any kind. In the set theory, the notions "set" and "element" are quite primitive in nature. For example, set A is defined as

$$A = \left[\text{red, man, tree, } 2, 5 \right]$$

with the bracketed terms as its elements. To indicate that "x is a member of set A", we write the following notation:

$$x \in A$$

Whenever x is not a member of A, we write

$$x \notin A$$

Generally, crisp sets are described in three ways:

1. **List method:** In this method, all elements of a set are defined by naming them. This method is suitable only for finite sets, where elements are limited in number. Set P, whose elements are p1, p2, p3 and p4, is written as

$$P = \{p1, p2, p3, p4\}$$

2. **Rule method:** In this case, a set is defined by a property satisfied by members of the set. This is represented as

$$M = \{x \mid F\}$$

which is read as M is a set with elements as x such that x has the property of F, where F is termed a "predicate of the set". For example, if Z represents all positive integers, then set T, comprising positive integers, is described as

$$T = \{x \mid x \in Z\}$$

3. **Characteristic function method:** In this method, a set is defined by a function termed a "characteristic function" denoted as "μ", which indicates which elements are members of the set or otherwise. Set A is defined by the characteristic function:

$$\mu_A(x) = \begin{cases} 1 & \text{for } x \in A \\ 0 & \text{for } x \notin A \end{cases}$$

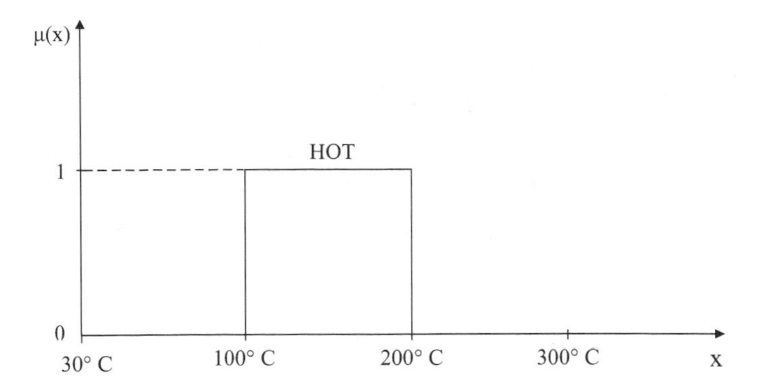

FIGURE 2.1 Representation of the characteristic function method.

The characteristic function method is represented in a diagrammatic form in Figure 2.1. In Figure 2.1, the variable is marked along the horizontal axis. The variable is assumed to be a continuous variable, say the temperature of a furnace. The crisp set is labelled HOT, defined as

$$\text{HOT} = \{x \mid 100°\text{C} \le x \le 200°\text{C}\}$$

This representation indicates that all elements of x from 100°C to 200°C have a μ(x) value of 1 and all remaining elements have 0 as their μ(x) value.

The classical set theory uses several operations, such as complement, intersection and difference. Let A and B be two classical sets in a universe U. The three important set operations can be defined as follows:

- Complement of A, $A' = \{x \mid x \notin A\}$
- Intersection of A and B, $A \cap B = \{x \mid x \in A \text{ and } x \in B\}$
- Union of A and B, $A \cup B = \{x \mid x \in A \text{ or } x \in B\}$

2.2.1 A Control Application Using Crisp Set

Consider a refrigerator motor that regulates the temperature of the inner chamber. The control logic is based on an ON-OFF control scheme. When the chamber temperature is below a set value, the motor is OFF and when the temperature increases above this value, the motor is switched ON to lower the temperature. Thus, the output is either "1" or "0", indicating that the motor is ON or OFF. This is primarily a crisp set operation. A motor runs at a rated speed when it is switched ON or else it does not run, i.e. zero-speed operation. The disadvantages of this method are a fluctuating dynamic response, non-uniform cooling and heavy overload at the time of starting the motor.

2.3 CONCEPT OF A FUZZY SET

As seen in Figure 2.1, a crisp set increases/decreases abruptly, making its elements totally disjoint with other members of the universe. However, such a strict categorisation does not exist where the human reasoning process is concerned. While observing a physical activity, we try to "model" it with different kinds of perceptions. Such a modelling process is the basic denominator for the analysis, design, simulation and implementation of the event. The model may be a simple statement, a mathematical expression, a figure, a block diagram, etc., and the variables of the model cannot be crisply defined. To illustrate this point, consider the following statement about the salary of Mr. X who works in a software firm.

Mr. X gets HIGH SALARY.

Let us attempt to use a crisp set to describe the concept of HIGH SALARY and assume that any salary above Rs.25,000 per month refers to HIGH SALARY. The contradiction here is that anyone who gets Rs.24,500 does not belong to HIGH SALARY. However, common sense does not permit such a dichotomy. Here, the crisp set concept is confronted with the fact that no demarcation exists between Rs.25,000 and Rs.24,500. Our concept about HIGH SALARY is a continuously varying surface, as indicated in Figure 2.2.

From Figure 2.2, the amount Rs.7,000 is definitely not true, while Rs.20,000 can be considered generally agreeable and Rs.40,000 is

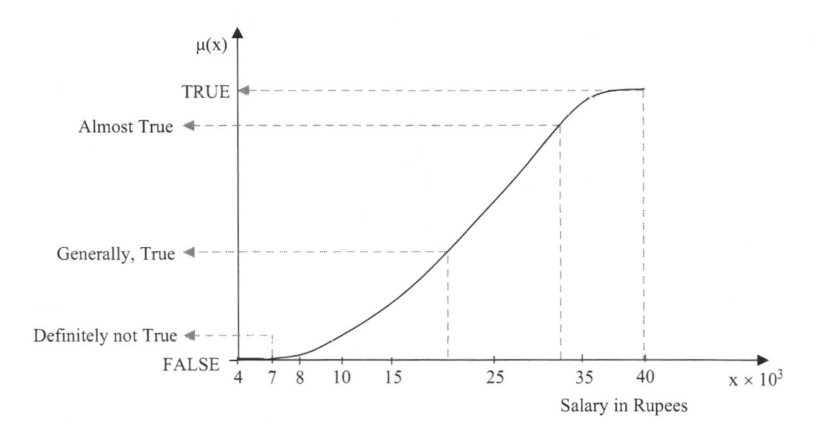

FIGURE 2.2 The fuzzy set concept.

definitely true with our concept about HIGH SALARY. Thus, as we proceed along the x-axis, the truth content about our concept, HIGH SALARY, becomes increasingly true. Therefore, it is obvious that the curve labelled HIGH SALARY is very compatible with our thinking process. This type of representation of a physical variable as a continuous curve is called a fuzzy set or membership function.

This concept of depicting continuous curves for a variable salary prompts one to think of the variable in terms other than HIGH SALARY. That is, the salary range can be further split into various labels. Another label to describe employees' salary can be LOW SALARY, which is shown in Figure 2.3(a).

Since HIGH SALARY increases as salary increases, LOW SALARY can be considered to decrease as salary increases. Such continuous curves, marked as LOW SALARY and HIGH SALARY, representing a variable with a varied degree of truth content are known as fuzzy sets: "LOW SALARY" and "HIGH SALARY". Thus, any variable can be represented as a continuous curve with a suitable label. The variable will have different levels of truth in each fuzzy set. Figure 2.3(a) shows that as salary increases along the x-axis, it becomes less and less LOW SALARY and more HIGH SALARY. However, it is also evident that LOW

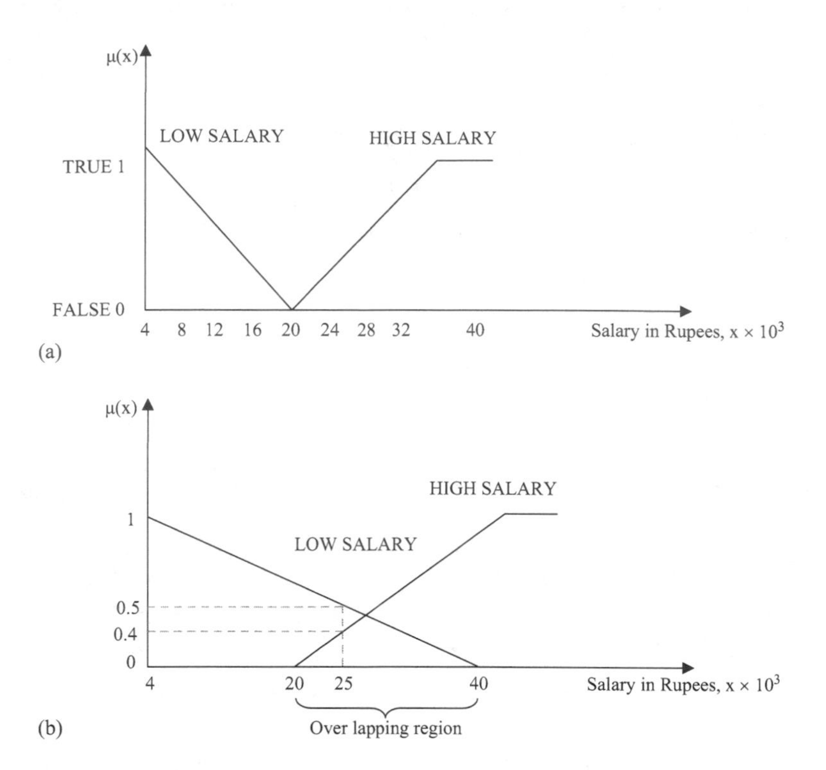

FIGURE 2.3 LOW SALARY and HIGH SALARY fuzzy sets (a) without overlap and (b) with overlap.

SALARY and HIGH SALARY encompass two crisp regions of salary. Hence, the two continuous curves must overlap to give a more logical and real-world representation of these two related concepts. This is shown in Figure 2.3(b).

From experience, we know that as salary increases, it goes from LOW SALARY to HIGH SALARY. Thus, salary has less truth value in LOW SALARY as it increases and gains more truth value in HIGH SALARY. The quantum of truth of salary in LOW SALARY is less when compared to that in HIGH SALARY. Now, it is possible to assign a numerical value between 0 and 1 for each value of salary, depending upon its compatibility with truth in each fuzzy set. This is shown in Figure 2.3(b).

The numerical value is termed "degree of membership" or "truth function" or more commonly "membership grade". This membership value makes a one-to-one correspondence between the element and the fuzzy set. This is indicated as $\mu_{LOW\ SALARY}(x)$ like a discriminant function. Thus, a salary of Rs.25,000 has a membership of 0.5 in LOW SALARY and 0.4 in HIGH SALARY. Hence, each variable has a certain degree of membership in each fuzzy set.

2.4 BASIC PROPERTIES AND CHARACTERISTICS OF FUZZY SETS

We will now examine a few of the important properties of fuzzy sets. A fuzzy set has several basic properties that affect the way the set is used and how it models a system.

2.4.1 Universe of Discourse

A system model has several inputs and outputs, each variable having its own minimum and maximum values. Thus, each variable is generally composed of multiple, overlapping fuzzy sets with each fuzzy set describing a semantic partition of the variable's total space. Figure 2.4 illustrates this concept. The model parameter SPEED is broken down into six fuzzy sets: ZERO, VERY SLOW, SLOW, MEDIUM, HIGH and VERY HIGH.

The total problem space from the minimum to the maximum allowable value of the variable under consideration is called the "universe of discourse". The universe of discourse for the model

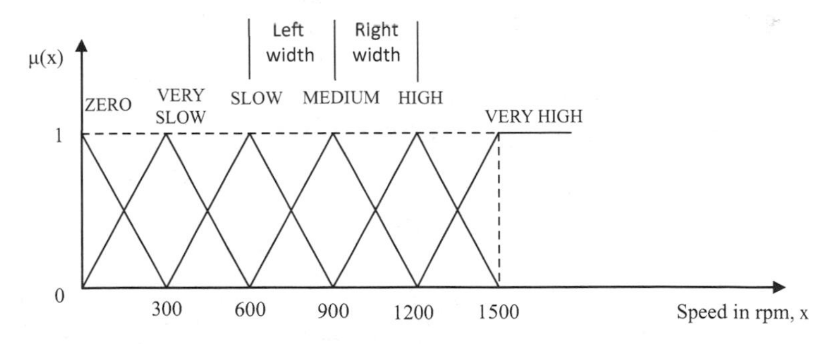

FIGURE 2.4 Fuzzy sets describing motor speed.

variable SPEED is 0 rpm to 1500 rpm. The universe of discourse is associated with a system variable and not with a particular fuzzy set.

2.4.2 Fuzzy Set Domain

A variable in a fuzzy set will have a minimum value and a maximum value with the two values encompassing all possible values in between. The total allowable problem space is divided into various fuzzy sets with each fuzzy set encompassing a specific range of values. The range of values covered by a particular fuzzy set is termed a "domain of the fuzzy set". The domain is a set of real numbers, increasing monotonically from left to right.

The values can be both positive and negative. The domain is selected to represent the complete operating range of values for the fuzzy sets within the context of the model. As an example, the fuzzy set SLOW has a domain from 300 rpm to 900 rpm, as shown in Figure 2.4, while the domain of the fuzzy set MEDIUM is from 600 rpm to 1200 rpm.

The domain has both an upper and a lower limit, even though, in reality, the scope of values associated with the fuzzy concept may be open-ended. For example, consider the fuzzy set VERY HIGH shown in Figure 2.4. The possible vehicle speed may go beyond 1500 rpm. However, as the fuzzy set reaches unity at 1500 rpm, any speed above this is definitely VERY HIGH.

2.4.3 Shouldered Fuzzy Set

Consider Figure 2.4 in which six fuzzy sets describe the variable speed. Here, the first and last fuzzy set, namely ZERO and VERY HIGH, are called shouldered fuzzy sets. The shape of shouldered fuzzy sets is different from the remaining fuzzy sets. These sets describe the initial and final domains in a universe of discourse.

2.4.4 Height of Fuzzy Set

This is the maximum membership grade of a fuzzy set. It is unity for all fuzzy sets in Figure 2.4. This aspect is further discussed in Section 3.3.3.3.

2.4.5 Overlap between Fuzzy Sets

Two adjacent fuzzy sets overlap each other with common elements in the overlapped domain. The quantum of overlap is less than 50% between fuzzy sets LOW SALARY and HIGH SALARY in Figure 2.3, whereas it is strictly 50% between all fuzzy sets in Figure 2.4.

2.4.6 Left and Right Width of Fuzzy Set

The domain of a fuzzy set can be divided on either side of the peak value of the fuzzy set: the left-side domain is termed “left width” and the right-side domain is termed “right width”. This is indicated for the fuzzy set MEDIUM in Figure 2.4.

2.4.7 Non-Convex and Convex Fuzzy Sets

Non-convex fuzzy sets are fuzzy sets in which the membership grade alternately increases and decreases in the domain.

Fuzzy sets in which membership grades do not alternately increase and decrease are called convex fuzzy sets. This means that the membership function of convex sets does not contain “dips”. The fuzzy sets in Figure 2.4 are convex fuzzy sets.

2.4.8 Commonly Employed Fuzzy Sets

A system variable can be suitably represented by the designer. Thus, fuzzy sets can take any random geometric shape. However, for ease of computation, certain standard fuzzy sets are available in the literature. For brevity, triangular and trapezoidal fuzzy sets are as follows:

Triangular fuzzy set

$$\Lambda(x;\alpha,\beta,\gamma)=\begin{cases}0, & x<\alpha\\ \left(\dfrac{x-\alpha}{\beta-\alpha}\right), & \alpha\le x<\beta\\ \left(\dfrac{\gamma-x}{\gamma-\beta}\right), & \beta\le x\le\gamma\\ 0, & x>\gamma\end{cases}$$

Trapezoidal fuzzy set

$$\Pi(x;\alpha,\beta,\gamma,\delta)=\begin{cases}0, & x<\alpha\\ \left(\dfrac{x-\alpha}{\beta-\alpha}\right), & \alpha\leq x<\beta\\ 1, & \beta\leq x<\gamma\\ \left(\dfrac{\delta-x}{\delta-\gamma}\right), & \gamma\leq x\leq\delta\\ 0, & x>\delta\end{cases}$$

These are shown in Figure 2.5(a,b).

2.5 CRISP FUZZY SET OPERATIONS

The classical set theory defines three major fundamental operations on sets, namely the complement, intersection and union operations.

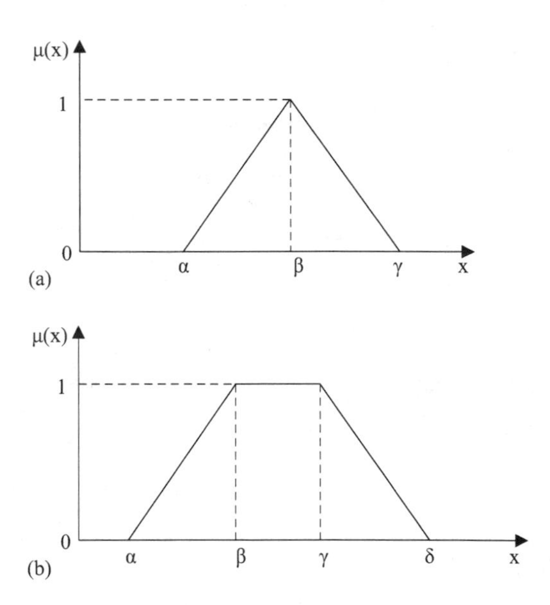

FIGURE 2.5 Representation of (a) triangular function and (b) trapezoidal function.

Let S_1 and S_2 be two classical sets in a universe U, which are represented in the list method as

$$S_1 = \{1, 2, \text{red}, \text{tree}, 7\} \text{ and}$$

$$S_2 = \{\text{mango}, \text{cow}, \text{red}, 2\}$$

The interaction of sets S_1 and S_2 is the set containing all the elements belonging to both set S_1 and set S_2. Thus,

$$S_1 \cap S_2 = \{2, \text{red}\}$$

The union of sets S_1 and S_2 contains all the elements that appear either in set S_1 or in set S_2 Thus,

$$S_1 \cup S_2 = \{1, 2, \text{red}, \text{tree}, 7, \text{mango}, \text{cow}\}$$

The complement of set S_1 consists of all the elements drawn from the possible universe of the set that are not in set S_1. The complement of S_1 is denoted $\sim S_1$.

There is an important distinction between fuzzy set logic and crisp set logic. While classical set membership "abruptly" changes, this is not the case with fuzzy sets. It is possible to redefine the set operations, namely union, intersection and complement, in terms of characteristic functions, which will be useful when dealing with fuzzy set operations. Let A and B be two classical sets and let $\mu_A(x)$ and $\mu_B(y)$ be their characteristic functions. This is given as

$$\mu_A(x) = \begin{cases} 0, & 0 \le x < x_1 \\ 1, & x_1 \le x \le x_n \\ 0, & x > x_n \end{cases}$$

$$\mu_B(y) = \begin{cases} 0, & 0 \le y < y_1 \\ 1, & y_1 \le y \le y_n \\ 0, & y > y_n \end{cases}$$

Two crisp sets are illustrated in Figures 2.6(a,b).

The three set operations – intersection, union and complement – are equivalent to the AND, OR and NOT operations of Boolean algebra, respectively.

- Intersection $\mu_A(x) \cap \mu_B(y) = \mu_A(x) \text{ AND } \mu_B(y)$
- Union $\mu_A(x) \cup \mu_B(y) = \mu_A(x) \text{ OR } \mu_B(y)$
- Complement $\sim \mu_A(x) = \overline{\mu_A(x)}$

These defined operations can be performed only if the variables x and y are specified. Here, the variables x and y are assigned a characteristic function and the set operations are performed on the characteristic function value rather than on the variables. Consider the variables x and y of sets A and B, respectively. The basic set operations performed on the characteristic function of these variables are as follows:

- Intersection

$$\mu_A(x = x_p) \cap \mu_B(y = y_q) = \mu_A(x_p) \text{ AND } \mu_B(y_q) = 0 \text{ AND } 1 = 0$$

- Union

$$\mu_A(x = x_p) \cup \mu_B(y = y_q) = \mu_A(x_p) \text{ OR } \mu_B(y_q) = 0 \text{ OR } 1 = 1$$

- Complement

$$\sim \mu_A(x = x_p) = \overline{\mu_A(x_p)} = \overline{0} = 1$$

The NOT operation is the only operation that can be performed on a single crisp set (say A or B alone). Figure 2.6(c) shows ~A.

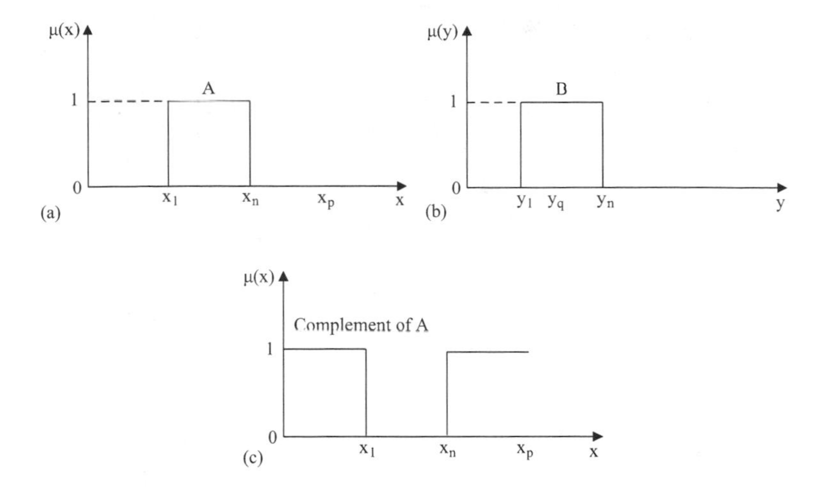

FIGURE 2.6 Representation of (a) crisp set A, (b) crisp set B and (c) complement of crisp set A.

2.6 BASIC OPERATIONS ON FUZZY SETS

Similar to conventional sets, there are specifically defined operations for combining and modifying fuzzy sets. These set theoretic functions provide the fundamental tools of the logic. The set operations – union, intersection and complement – are defined in the exact same way as for classical sets in terms of characteristic function. Consider two fuzzy sets A and B, as shown in Figures 2.7(a,b), with membership functions $\mu_A(x_p)$ and $\mu_B(y_q)$, respectively. These two fuzzy sets can be combined in different ways as follows:

- Intersection $\mu_A(x) \cap \mu_B(y) = \min\{\mu_A(x), \mu_B(y)\}$
- Union $\mu_A(x) \cup \mu_B(y) = \max\{\mu_A(x), \mu_B(y)\}$
- Complement $\sim \mu_A(x) = 1 - \mu_A(x)$

Since fuzzy sets are not crisply partitioned in the same way as Boolean sets, these operations are applied on the truth content of fuzzy sets.

In fuzzy logic, the intersection of two fuzzy sets A and B with membership functions $\mu_A(x_p)$ and $\mu_B(y_q)$ is defined as the minimum of the two individual membership functions. The intersection operation is the equivalent of the AND operation in Boolean algebra. Considering fuzzy sets A and B in Figures 2.7(a,b), the intersection is defined as

$$\mu_A(x = x_1) \cap \mu_B(y = y_1) = \min\{\mu_A(x_1), \mu_B(y_1)\}$$
$$= \min\{0.3, 0.4\} = 0.3$$

$$\mu_A(x = x_2) \cap \mu_B(y = y_3) = \min\{\mu_A(x_2), \mu_B(y_3)\}$$
$$= \min\{0.8, 0\} = 0$$

The union of two sets is equivalent to the logical OR operation and is performed by taking the maximum of the truth membership values. For the given fuzzy sets A and B, the union is

$$\mu_A(x = x_1) \cup \mu_B(y = y_1) = \max\{\mu_A(x_1), \mu_B(y_1)\}$$
$$= \max\{0.3, 0.4\} = 0.4$$

$$\mu_A(x = x_2) \cup \mu_B(y = y_3) = \max\{\mu_A(x_2), \mu_B(y_3)\}$$
$$= \max\{0.8, 0\} = 0.8$$

The complement of fuzzy set A with membership function $\mu_A(x)$ is defined as the negation of the specified membership function and is the equivalent of the NOT operation in Boolean algebra. Given a fuzzy set A, its complement is such that

$$\sim \mu_A(x = x_1) = 1 - \mu_A(x_1) = 1 - 0.3 = 0.7$$

Figure 2.7(c) shows the complement of fuzzy set A.

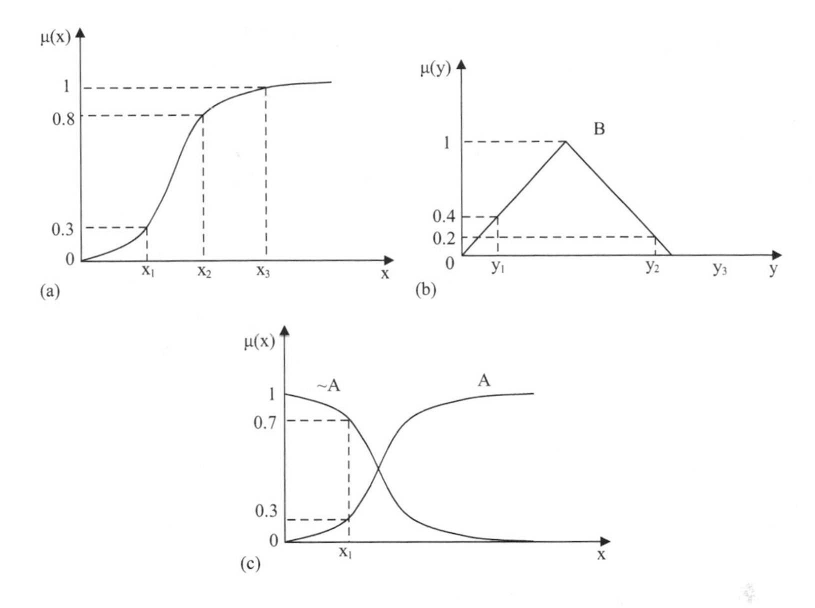

FIGURE 2.7 Fuzzy representations of (a) set A, (b) set B and (c) complement of set A.

2.6.1 Intersection of Fuzzy Sets

In a crisp system, the intersection of two sets contains the elements that are common to both sets. This is also equivalent to the logical AND operation. In fuzzy logic, the AND operator is performed by taking the minimum of the truth membership grades. The intersection operator is the most commonly used operator in a fuzzy rule base. To illustrate the fuzzy AND operation and also to compare its results with crisp ANDing, let us consider the following examples:

Example 1

A company intends to recruit a reasonably experienced (but not old) person for its R&D wing. The recruit should also have an excellent overall cumulative grade point average (CGPA). Thus, the person should be middle-aged with a very good academic record. The characteristic function for MIDDLE AGED is given by

$$\mu_{\text{MIDDLE AGED}}(x) = \begin{cases} 1, & 25 \leq x \leq 35 \\ 0, & \text{Otherwise} \end{cases}$$

Thus, the members of the set are all individuals between the ages of 25 and 35, both inclusive. The membership graph for this crisp set is shown in Figure 2.8(a).

The characteristic function for VERY GOOD is given by

$$\mu_{\text{VERY GOOD}}(y) = \begin{cases} 1, & \text{for } y \geq 0.8 \\ 0, & \text{Otherwise} \end{cases}$$

Thus, all individuals whose CGPA is above 8.0 are VERY GOOD. This is represented in Figure 2.8(b).

In Boolean logic, those individuals who are both middle-aged and have a higher grade are found with the AND operator. This is equivalent to the bitwise ANDing of Boolean vectors representing the truth of the set characteristic expression for each category. Since these are Boolean discriminants, the results can only be true [1] or false [0]. Let there be seven candidates, namely Akshaya, Raju, etc. The age and CGPA of each individual are given in Table 2.1. The characteristic function values of the age and CGPA of each individual are obtained from the crisp sets MIDDLE AGED and VERY GOOD. Now, the Boolean AND operation is performed and the results are shown in Table 2.1.

Thus, as per this Boolean AND process, two people are qualified for the post. The other candidates are rejected straightaway and are out as per this logic.

It is interesting to get the solution using the fuzzy AND operation. For this, we have to construct fuzzy membership functions depicting the foregoing variables. In a fuzzy set, the membership edge is not abruptly disjoint at a particular domain value, but reflects the degree of membership for

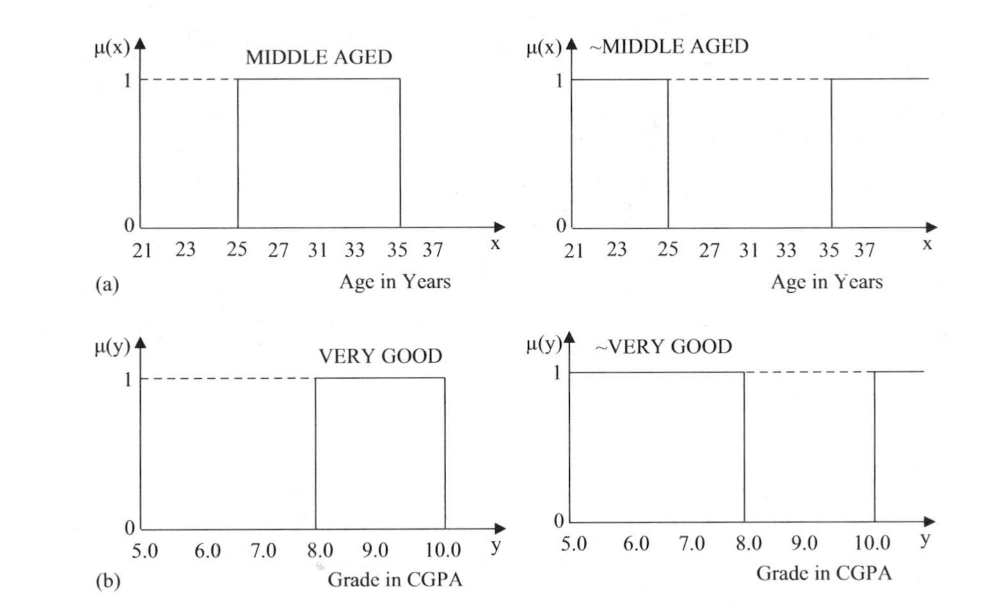

FIGURE 2.8 Crisp set representations: (a) MIDDLE AGED and its complement and (b) VERY GOOD and its complement.

TABLE 2.1 Crisp AND Operation of MIDDLE AGED and VERY GOOD

			Crisp Set Operation		
Name	Age	Grade	$\mu_{MIDDLE\ AGED}(x)$	$\mu_{VERY\ GOOD}(y)$	$\mu_{MIDDLE\ AGED}(x)$ AND $\mu_{VERY\ GOOD}(y)$
Akshaya	28	9.5	1	1	1
Raju	27	6.2	1	0	0
Binu	45	4.0	0	0	0
Tanuj	30	7.6	1	0	0
Venkat	23	8.0	0	1	0
Priya	28	8.5	1	1	1
Harini	36	5.0	0	0	0

values that lie at the extremes. Figure 2.9(a) is a fuzzy set representing MIDDLE AGED. This starts at 20 as the youngest age that someone might have the smallest degree of middle age. The membership curve climbs steadily until it reaches 30 years old, the absolute quintessence of middle age. After 30, the curve drops off again, so that someone around 35 is only moderately middle-aged and someone over 40 has no membership among the middle-aged.

The same kind of approach is applied in Figure 2.9(b) to the concept of VERY GOOD. For simplicity, the fuzzy set is represented as a straight line. As grade increases, membership in VERY GOOD increases proportionally. The fuzzy set starts at 5 as the value with zero membership in the set and increases until unity at 10 is reached. All grades below 5 are absolutely NOT FIT and all grades above 5 are relatively FIT.

Following the basic Zadeh rules for fuzzy intersection, the common space between the two sets is found by applying the operation:

$$\mu_A(x) \cap \mu_B(y) = \min\{\mu_A(x), \mu_B(y)\}$$

Table 2.2 shows the results of the intersection of the fuzzy sets MIDDLE AGED and VERY GOOD for each instance

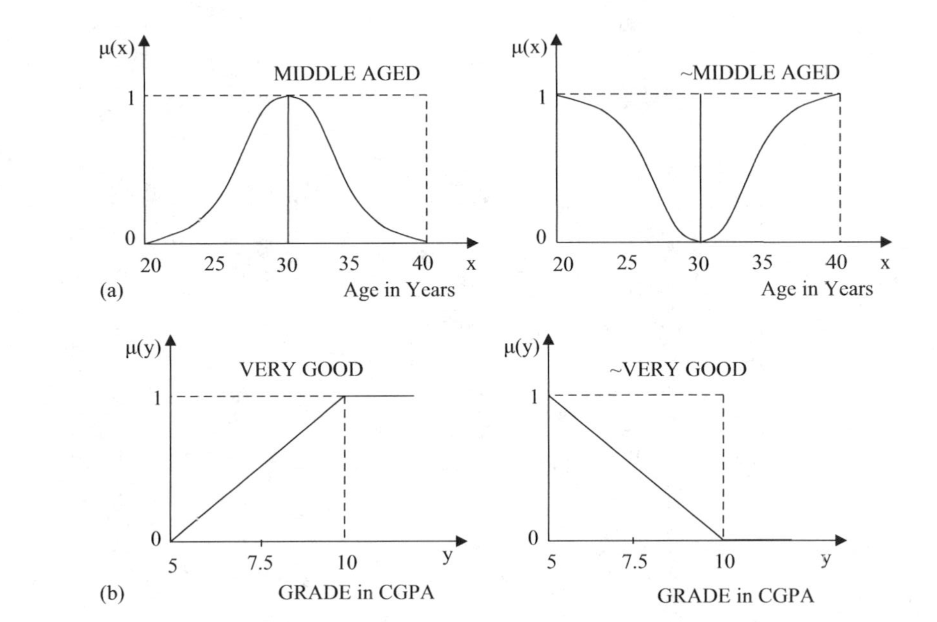

FIGURE 2.9 Fuzzy set representations of (a) MIDDLE AGED and its complement and (b) VERY GOOD and its complement.

TABLE 2.2 Fuzzy AND Truth Vectors for MIDDLE AGED and VERY GOOD

			Crisp Set Operation		
Name	**Age**	**Grade**	$\mu_{MIDDLE\ AGED}(x)$	$\mu_{VERY\ GOOD}(y)$	$\mu_{MIDDLE\ AGED}(x)$ **AND** $\mu_{VERY\ GOOD}(y)$
Akshaya	28	9.5	0.92	0.9	0.9
Raju	27	6.2	0.48	0.24	0.24
Binu	45	4.0	0	0	0
Tanuj	30	7.6	1	0.52	0.52
Venkat	23	8.0	0.3	0.6	0.3
Priya	28	8.5	0.9	0.7	0.7
Harini	36	5.0	0.3	0	0

in the sample population. The last column shows the truth value resulting from the intersection or ANDing of the two truth values for MIDDLE AGED and VERY GOOD.

In contrast to the Boolean intersection, the fuzzy operation identifies five instances with different degrees of compatibility with the concept of MIDDLE AGED and VERY GOOD. In particular, we can see that Raju, Tanuj and Venkat have been included in the selection, corresponding to the common-sense view that, although they fall outside a strict Boolean partition, they are semantically close to the concept we enforced (as opposed to enforcing a rigid arithmetic rule). Further comparison of Tables 2.1 and 2.2 indicates that both methods select Akshaya and Priya as the "best" candidates. Thus, the fuzzy AND operation fully endorses the Boolean AND operation. In addition, it gives a solution space with a varied degree of compatibility to the concept of VERY GOOD and MIDDLE AGED.

Example 2

Consider the case of a person who wants to purchase a house. Let the house he wants to purchase be VERY BIG AND CHEAP. The variables that are used to define the sets

VERY BIG and CHEAP are the area of the house in square feet and the cost of the house in lakhs of rupees, respectively, and the variables are context dependent. Let the characteristic function for the crisp set VERY BIG be defined as

$$\mu_{\text{VERY BIG}}(x) = \begin{cases} 1, & \text{for } x \geq 2500 \text{ sq ft} \\ 0, & \text{Otherwise} \end{cases}$$

Thus, the members of the set VERY BIG are houses with an area greater than or equal to 2500 sq ft. This is represented in Figure 2.10(a).

Similarly, the characteristic function set CHEAP is given by

$$\mu_{\text{CHEAP}}(y) = \begin{cases} 1, & \text{for } y \leq 5.0 \text{ lakhs} \\ 0, & \text{Otherwise} \end{cases}$$

This set specifies the members of the set CHEAP as houses that cost less than or equal to Rs.5 lakhs. This is represented in Figure 2.10(b).

To find a solution to the problem stated, the AND operator is used. Table 2.3 shows the results of applying the Boolean AND operator to a sample of houses.

From the results, it can be interpreted that only house H5 suits the requirement as per the crisp set operation and the other choices are rejected outright.

For the same problem, the solution can be obtained using the fuzzy AND operation. Figure 2.11(a) represents the fuzzy set VERY BIG. This starts at an area of 800 sq ft with a membership value of 0.09 and the membership curve climbs steadily until it reaches 6000 sq ft with a membership value of 1.

Similarly, the fuzzy set CHEAP is represented as in Figure 2.11(b). The fuzzy set CHEAP starts at Rs.2 lakhs with a membership value of 1 and decreases proportionally until it

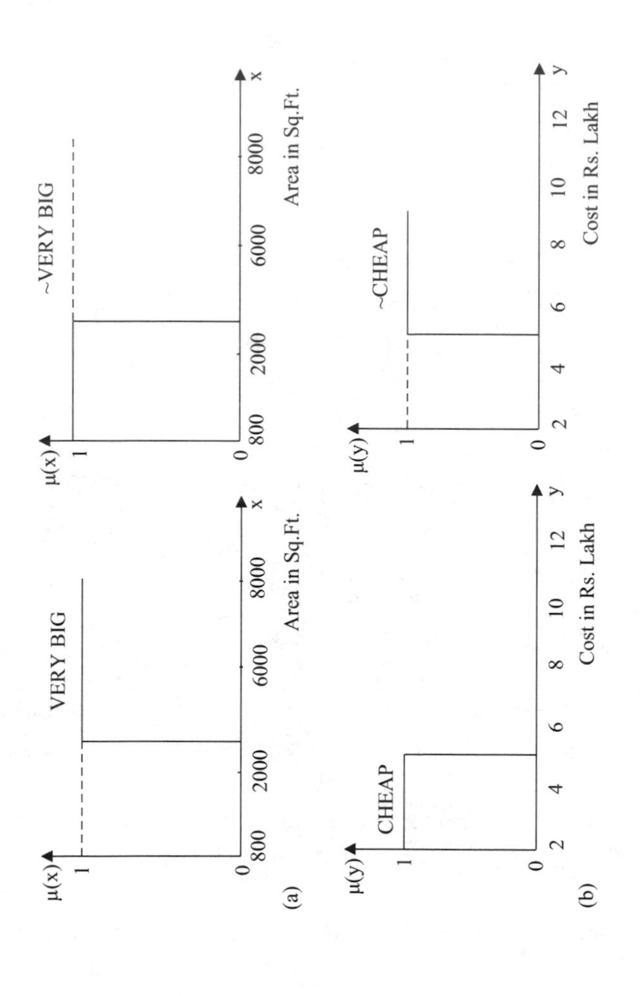

FIGURE 2.10 Crisp set representations of (a) VERY BIG and its complement and (b) CHEAP and its complement.

TABLE 2.3 Crisp AND Operation of VERY BIG and CHEAP

House	Area (sq ft)	Cost in Rs. Lakh	Crisp Set Operation		
			$\mu_{VERY\ BIG}(x)$	$\mu_{CHEAP}(y)$	$\mu_{VERY\ BIG}(x)$ AND $\mu_{CHEAP}(y)$
H1	2000	3.00	0	1	0
H2	4500	6.70	1	0	0
H3	7000	11.00	1	0	0
H4	800	1.80	0	1	0
H5	2800	4.30	1	1	1

reaches Rs.10 lakhs with a membership value of zero. The cost of a house less than Rs.10 lakhs is relatively CHEAP and that greater than Rs.10 lakhs is not CHEAP.

Using the fuzzy AND operation, the solution can be obtained by taking the minimum of the two membership functions for the two sets. That is

$$\mu_{VERY\ BIG}(x) \cap \mu_{CHEAP}(y) = \min\{\mu_{VERY\ BIG}(x), \mu_{CHEAP}(y)\}$$

Table 2.4 shows the result of the fuzzy AND operation. The fuzzy operation gives maximum weightage to house H5 as in the case of the crisp operation. However, in the crisp operation, there is only one choice and it gives a demarcated boundary between the variables. On the other hand, there are several choices with the fuzzy AND operation. One must remember that purchasing a house is an important event in one's life. Though the cost and area of a house are important, there are several other factors such as the house's surroundings, the availability of essential services such as transport, banking and hospital facilities, the climatic conditions at the location and the cost of living also influence the decision to purchase a house. In this context, the crisp set solution is mostly narrow. Rather than giving a single solution, the fuzzy ANDing operation provides a solution space encompassing all the variables with variable grades. The output of the crisp operation is a firm yes or a firm no, whereas the

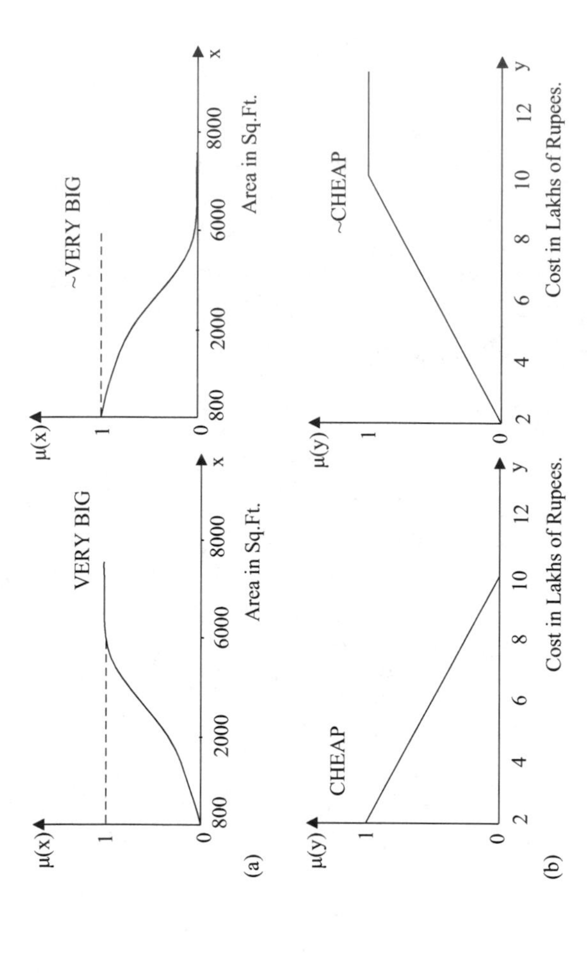

FIGURE 2.11 Fuzzy set representation of (a) VERY BIG and its complement and (b) CHEAP and its complement.

TABLE 2.4 Fuzzy AND Truth Vectors for VERY BIG and CHEAP

House	Area (sq ft)	Cost in Rs. Lakh	Fuzzy Set Operation		
			$\mu_{VERY\ BIG}(x)$	$\mu_{CHEAP}(y)$	$\mu_{VERY\ BIG}(x)$ AND $\mu_{CHEAP}(y)$
H1	2000	3.00	0.3	0.9	0.3
H2	4500	6.70	0.75	0.4	0.4
H3	7000	11.00	1	0	0
H4	800	2.00	0.09	1	0.09
H5	2800	4.30	0.6	0.75	0.6

fuzzy output provides a degree of how favourable it would be for the person to buy the house.

2.6.2 Union of Fuzzy Sets

The union of two sets finds those elements that are members in both sets and the union is performed with the OR operator. In fuzzy logic, the OR operator is supported by taking the maximum of the truth membership grades. The fuzzy OR operator is less frequently used in systems since only one proposition of the largest membership alone is considered while all other propositions are discarded. However, there are several cases where one may opt for a larger advantage among various choices. Consider a few commonly heard statements:

- I will opt for an engineering course in a reputed institute now OR try for an MBA later.
- After retirement, I may settle in my native place OR in Bangalore.
- During this summer vacation, I am planning to go to Ooty OR Kodaikanal.

In order to illustrate these points more vividly, consider the previous two examples.

Example 1

Using the same problem of finding middle-aged candidates with higher grades, we can examine the effects of the union operator on crisp and fuzzy sets. Here, we want to find members of the population who are either middle-aged or have a higher grade similar to the AND process. We can visualise this process as the ORing of bit vectors representing the truth of the characteristic function at each instance in the population. Table 2.5 shows the results of applying the Boolean OR operator to the population of candidates.

The OR operator provides the least amount of discrimination in the sets and selects any case where either of the discriminant functions is true. Using the fuzzy sets in Example 1 in Section 2.6.1.1 on fuzzy intersection, the union of VERY GOOD and MIDDLE AGED in a fuzzy system can also be determined. Following the Zadeh rules for fuzzy union, the common conjoined space between the two sets is found by applying the operation:

$$\mu_A(x) \cup \mu_B(y) = \max\{\mu_A(x), \mu_B(y)\}$$

Table 2.6 shows the results of a union of the fuzzy sets MIDDLE AGED and VERY GOOD for each instance in our sample population.

The comparison of a crisp set and a fuzzy set operation shows that while the crisp OR operation "blindly" selects

TABLE 2.5 Crisp OR Operation of MIDDLE AGED and VERY GOOD

			Crisp Set Operation		
Name	**Age**	**Grade**	$\mu_{MIDDLE\ AGED}(x)$	$\mu_{VERY\ GOOD}(y)$	$\mu_{MIDDLE\ AGED}(x)$ **AND** $\mu_{VERY\ GOOD}(y)$
Akshaya	28	9.5	1	1	1
Raju	27	6.2	1	0	1
Binu	45	4.0	0	0	0
Tanuj	30	7.6	1	0	1
Venkat	23	8.0	0	1	1
Priya	28	8.5	1	1	1
Harini	36	5.0	0	0	0

TABLE 2.6 Fuzzy AND Truth Vectors for MIDDLE AGED and VERY GOOD

Name	Age	Grade	Crisp Set Operation		
			$\mu_{MIDDLE\ AGED}(x)$	$\mu_{VERY\ GOOD}(y)$	$\mu_{MIDDLE\ AGED}(x)$ AND $\mu_{VERY\ GOOD}(y)$
Akshaya	28	9.5	0.92	0.9	0.92
Raju	27	6.2	0.48	0.24	0.48
Binu	45	4.0	0	0	0
Tanuj	30	7.6	1	0.52	1
Venkat	23	8.0	0.3	0.6	0.6
Priya	28	8.5	0.98	0.7	0.98
Harini	36	5.0	0.7	0	0.7

five people, the fuzzy OR operation gives a selection list with different grading values, which helps to distinguish between different candidate solutions.

Example 2

To examine the effects of the union operator on crisp and fuzzy sets, the problem of purchasing a VERY BIG and CHEAP house can be stated as follows:

The person needs either a VERY BIG house OR a CHEAP house.

To find a solution to the problem stated, the OR operator can be used. Table 2.7 shows the results of applying the Boolcan OR operator to a sample of houses.

TABLE 2.7 Crisp OR Truth Vectors for VERY BIG and CHEAP

House	Area (sq ft)	Cost in Rs. Lakh	Crisp Set Operation		
			$\mu_{VERY\ BIG}(x)$	$\mu_{CHEAP}(y)$	$\mu_{VERY\ BIG}(x)$ AND $\mu_{CHEAP}(y)$
H1	2000	3.00	0	1	1
H2	4500	6.70	1	0	1
H3	7000	11.00	1	0	1
H4	800	1.80	0	1	1
H5	2800	4.30	1	1	1

TABLE 2.8 Fuzzy OR Truth Vectors for VERY BIG and CHEAP

House	Area (sq ft)	Cost in Rs. Lakh	Fuzzy Set Operation		
			$\mu_{VERY\ BIG}(x)$	$\mu_{CHEAP}(y)$	$\mu_{VERY\ BIG}(x)$ AND $\mu_{CHEAP}(y)$
H1	2000	3.00	0.3	0.9	0.9
H2	4500	6.70	0.75	0.4	0.75
H3	7000	11.00	1	0	1
H4	800	1.80	0.09	1	1
H5	2800	4.30	0.6	0.75	0.75

Thus, as per this Boolean OR operation, all houses are equally valid. The solution to the problem can also be obtained using the fuzzy OR operation by taking the maximum of the two membership functions for the two sets VERY BIG and CHEAP.

$$\mu_{VERY\ BIG}(x) \cup \mu_{CHEAP}(y) = \max\{\mu_{VERY\ BIG}(x), \mu_{CHEAP}(y)\}$$

Table 2.8 shows the result of the fuzzy OR operation.

The fuzzy operation highlights houses H3 and H4 as the best options. As per the crisp operation, all the houses are equally valid and choices are ruled out. The crisp operation takes care of constraints, whereas the fuzzy operation fully endorses the Boolean operation. In addition, it grades various options.

2.6.3 Complement (Negation) of Fuzzy Sets

A complement or negation of a set A contains all the elements that are not in A and the complement is represented by the characteristic function:

$$\sim \mu_A(x) = 0 \quad \text{if } x \in A$$
$$= 1 \quad \text{if } x \notin A$$

To see how the complement of a fuzzy region differs significantly from its Boolean counterpart, let us consider the sets MIDDLE AGED and VERY GOOD in Example 1 of Section 2.6.1.1 discussed in the fuzzy intersection and union operation. Let us find the variables that are neither MIDDLE AGED nor VERY GOOD. The characteristic function for the crisp sets ~MIDDLE AGED and ~VERY GOOD are given by

$$\sim\mu_{\text{MIDDLE AGED}}(x) = \begin{cases} 0, & 25 \le x \le 35 \\ 1, & \text{Otherwise} \end{cases}$$

$$\sim\mu_{\text{VERY GOOD}}(y) = \begin{cases} 0, & y \ge 8.0 \\ 1, & \text{Otherwise} \end{cases}$$

The complement of the crisp sets MIDDLE AGED and VERY GOOD appears as in Figures 2.8(a,b), respectively.

To answer the question of which members of the population are not VERY GOOD and not MIDDLE AGED, we plot the complements of fuzzy sets MIDDLE AGED and VERY GOOD as shown in Figure 2.9(a,b).

In fuzzy set logic, the complement is produced by inverting the truth function along each point of the fuzzy set. This is done by applying the law:

$$\sim \mu_A(x) = 1 - \mu_A(x)$$

TABLE 2.9 AND Operation of Crisp Sets ~MIDDLE AGED and ~VERY GOOD

			Crisp Set Operation		
Name	**Age**	**Grade**	$\sim\mu_{\text{MIDDLE AGED}}(x)$	$\sim\mu_{\text{VERY GOOD}}(y)$	$\sim\mu_{\text{MIDDLE AGED}}(x)$ **AND** $\sim\mu_{\text{VERY GOOD}}(y)$
Akshaya	28	9.5	0	0	0
Raju	27	6.2	0	1	0
Binu	45	4.0	1	1	1
Tanuj	30	7.6	0	1	0
Venkat	23	8.0	1	0	0
Priya	28	8.5	0	0	0
Harini	36	5.0	1	1	1

Table 2.9 shows the results of applying the intersection operator to the complement of the fuzzy sets MIDDLE AGED and VERY GOOD.

Similar to the intersection and union operations, a fuzzy complement does not dichotomise the set into crisply defined partitions. The complement registers the degree to which an element is complementary to the underlying fuzzy set concept. That is, how compatible is an element's value [X] with the assertion, X is NOT Y, where X is an element from the domain and Y is a fuzzy region. We can see this clearly by restating the same intersection of ~MIDDLE AGED and ~VERY GOOD in fuzzy form. The membership is generated by applying the discriminant function:

$$\sim \mu_{\text{MIDDLE AGED}}(x) \cap \sim \mu_{\text{VERY GOOD}}(y)$$

$$= \min\left(\sim \mu_{\text{MIDDLE AGED}}(x), \sim \mu_{\text{VERY GOOD}}(y)\right)$$

The results are given in Table 2.10.

Now consider Example 2 in Section 2.6.1.2 discussed in fuzzy intersection and union operation. Here, for the complement operation the problem is redefined as:

The person neither needs a VERY BIG house nor a CHEAP house.

Figure 2.10(a,b) illustrates the complement of crisp sets VERY BIG and CHEAP.

In fuzzy logic, the complement of sets VERY BIG and CHEAP is represented in Figure 2.11(a,b).

Tables 2.11 and 2.12 show the results of applying the intersection operator to the complement of crisp and fuzzy sets VERY BIG and CHEAP.

From the results, it can be seen that no house satisfies the person's requirement in the crisp set operation, whereas there are three choices in the fuzzy set operation. Here too, the superiority of the fuzzy operation is evident over the crisp operation.

TABLE 2.10 AND Vectors for Fuzzy Sets ~MIDDLE AGED and ~VERY GOOD

			Crisp Set Operation		
Name	**Age**	**Grade**	$\sim\mu_{MIDDLE\ AGED}(x)$	$\sim\mu_{VERY\ GOOD}(y)$	$\sim\mu_{MIDDLE\ AGED}(x) \cap \sim\mu_{VERY\ GOOD}(y)$
Akshaya	28	9.5	0.08	0.1	0.08
Raju	27	6.2	0.52	0.76	0.52
Binu	45	4.0	1.0	1.0	1.0
Tanuj	30	7.6	0.0	0.48	0
Venkat	23	8.0	0.7	0.4	0.4
Priya	28	8.5	0.02	0.3	0.02
Harini	36	5.0	0.3	1.0	0.3

The complement operation is employed where the solution space consists of a greater number of elements that do not satisfy most of the desired constraints. For example, let there be a pool of solution space with "n" number of candidates. A company wants to recruit candidates with good grades and good communication skills. If most of the students who appeared for selection do not satisfy this requirement, then perform a complement operation to highlight the candidates who satisfy this requirement.

Consider another example where a fault in a printed circuit board needs to be detected. To detect one or two damaged integrated circuits (ICs) instead of checking all the ICs in the printed circuit board containing many components, say 30 ICs connected together, the complement operation can be performed to highlight

TABLE 2.11 Crisp AND Operation of ~VERY BIG and ~CHEAP

			Crisp Set Operation		
House	**Area (sq ft)**	**Cost in Rs. Lakh**	$\sim\mu_{VERY\ BIG}(x)$	$\sim\mu_{CHEAP}(y)$	$\sim\mu_{VERY\ BIG}(x)$ AND $\sim\mu_{CHEAP}(y)$
H1	2000	3.00	1	0	0
H2	4500	6.70	0	1	0
H3	7000	11.00	0	1	0
H4	800	1.80	1	0	0
H5	2800	4.30	0	0	0

TABLE 2.12 The Fuzzy AND Truth Vectors for ~VERY BIG and ~CHEAP

House	Area (sq ft)	Cost in Rs. Lakh	Crisp Set Operation		
			$\sim\mu_{VERY\ BIG}(x)$	$\sim\mu_{CHEAP}(y)$	$\sim\mu_{VERY\ BIG}(x)$ AND $\sim\mu_{CHEAP}(y)$
H1	2000	3.00	0.7	0.1	0.1
H2	4500	6.70	0.25	0.6	0.25
H3	7000	11.00	0	1	0
H4	800	1.80	0.91	0	0
H5	2800	4.30	0.4	0.25	0.25

the damaged ICs. So, the complement operation can be employed where the solution space contains plenty of undesired elements.

2.7 CONCLUSION

The basic definition and characteristics of a fuzzy set have been well explained in this chapter. Commonly employed fuzzy sets have been included for completeness. Further, various Zadeh operations on fuzzy sets have been exemplified and the results compared with corresponding crisp set operations.

QUESTIONS

1. Explain three different ways of representing crisp sets.
2. If A = {1,7,9}, find ~A.
3. Give an example of a crisp set.
4. Sketch a crisp set describing distance.
5. Sketch a crisp set on temperature.
6. If A = {1,2} and B = {4,2}, find A ∩ B.
7. If A = {2,mango} and B = {3,2}, find A ∩ B.
8. Explain the concept of fuzzy set using a simple example.
9. Illustrate the concept of developing a fuzzy set labelled as follows: "PERFORMANCE" of Mr. X in subject Y is very good.

10. Sketch a fuzzy set describing temperature as the variable.
11. Define and explain with sketches the terms: (i) universe of discourse; (ii) domain of a fuzzy set; (iii) peak value of a fuzzy set; (iv) overlap between fuzzy sets; and (v) convex fuzzy sets.
12. What is a shouldered fuzzy set?
13. Sketch two fuzzy sets with mathematical equations.
14. Simple fuzzy sets are generally preferred. Why?
15. Why are only standard fuzzy sets recommended for fuzzy systems?
16. What is the significance of negated fuzzy sets? Justify.
17. Sketch the negated fuzzy sets of triangular and trapezoidal sets.
18. Sketch two fuzzy sets with 10% overlap.
19. A commercial three-phase induction motor, rated 440 V, 50 Hz, 7.5 A and 1435 rpm, is supplied always at a rated voltage and a variable frequency. Develop fuzzy sets to describe motor speed as the variable when the frequency varies from 0 to 100 Hz during no-load.
20. The ratings of a commercially available three-phase induction motor are given as 440 V, 7.5 A, 1440 rpm and 3.7 kW. It is expected that the load on the shaft varies from no-load to rated value. Develop fuzzy sets to describe the motor current and speed.
21. LOW, MED and HIGH are three adjacent fuzzy sets and are described as

$$\mathrm{LOW}(x) = \frac{-x}{30} + 1, \quad x \leq 30$$

$$\text{LOW}(x) = 0, \quad x > 30$$

$$\text{MED}(x) = \frac{1}{10}(x - 10), \quad 10 \le x \le 20$$

$$\text{MED}(x) = \frac{-1}{10}(x - 20) + 1, \quad 20 \le x \le 30$$

$$\text{HIGH}(x) = \frac{1}{10}(x - 20), \quad 20 \le x \le 30$$

$$\text{HIGH}(x) = 1, \quad x > 30$$

Sketch the fuzzy sets.

22. The cost x (in lakhs) of a 1000 sq ft house decreases as its distance y (in km) from the city centre increases. The fuzzy sets LESS and MORE describe x and y, respectively, and are given as

$$\text{MORE}(y) = \frac{1}{50}y, \quad 0 \le y \le 50$$

$$\text{LESS}(x) = \frac{-1}{50}(x - 10) + 1, \quad 10 \le x \le 60$$

Sketch the fuzzy sets.

23. A lathe machine uses reversible speed with maximum amplitudes from 300 rpm to 1500 rpm in opposite directions. Develop seven fuzzy sets with suitable names to describe the speed. Out of the seven sets, two fuzzy sets must be shouldered fuzzy sets and the remaining fuzzy sets should be similar in shape with 50% overlap.

24. Sketch a fuzzy set described as follows:

$$f(x;10,20,30) = \begin{cases} 0 & x \le 10 \\ 2\left(\dfrac{x-10}{20}\right)^2 & 0 < x \le 20 \\ 1-2\left(\dfrac{x-30}{20}\right)^2 & 20 < x \le 30 \\ 1 & x > 30 \end{cases}$$

25. Define the fuzzy AND operation.
26. If $\mu_A\ (x = 0.3)$ and $\mu_B\ (y = 0.055)$, find $\mu_A(x) \cap \mu_B(y)$.
27. Illustrate the fuzzy AND operation with an example.
28. Illustrate the superiority of the fuzzy AND operation over the crisp AND operation with a suitable example.
29. You are planning to purchase a TV and you have received a few offers of TVs from different companies. Frame this as an AND operation and obtain the best TV using the fuzzy AND method. Also, compare the solution with that obtained through the crisp AND method.
30. Compare and understand the fuzzy and crisp AND operations by considering the selection of students to a post-graduate programme in a reputed institute.
31. A financial firm is interested in recruiting security staff. Illustrate how the fuzzy AND operation can be used for this purpose. Also compare the result with the crisp AND operation.
32. It is proposed to recruit security staff to guard your old farmhouse in your village. Illustrate the selection as a fuzzy AND operation.
33. Define the fuzzy OR operation.

34. If A={4, mango} and B={7, cow}, find A ∪ B.

35. If $\mu_A\ (x = 0.5)$ and $\mu_B\ (y = 0.31)$, find $\mu_A(x) \cup \mu_B(y)$.

36. What is meant by COMPLEMENT of a fuzzy set? Give a typical application of the same.

CHAPTER 3

Fuzzy Reasoning

3.1 INTRODUCTION

This chapter explains the various internal operations involved in a fuzzy logic system, including fuzzification, the inference engine mechanism, different implication methods and popular defuzzification methods. Different aspects governing the choice of defuzzification schemes are also outlined.

3.2 A CONVENTIONAL CONTROL SYSTEM

A better understanding of a fuzzy logic system can be gained if we first deal with a conventional control system block diagram. A closed loop control system is shown in Figure 3.1.

Here, the task is to maintain the plant output at the desired level, ω_r^*, in spite of disturbances caused to the plant. In order to obtain the best dynamic response at each operating point of the plant, a suitable controller needs to be designed. Conventionally, there are three types of controllers, namely proportional (P), proportional-integral (PI) and proportional-integral-derivative (PID). These controllers are characterised by the following equations:

For the P controller:

$$m(t) = K_p c(t) \tag{3.1}$$

For the PI controller:

$$m(t) = K_p e(t) + K_i \int_0^t e(t)\,dt \tag{3.2}$$

For the PID controller:

$$m(t) = K_p e(t) + K_i \int_0^t e(t)\,dt + K_d \frac{de(t)}{dt} \tag{3.3}$$

In these equations, K_p, K_i and K_d are known as proportional, integral and derivative constants. Depending on the stability requirements of the system and the desired dynamic response, a P, PI or PID controller can be selected. In most cases, a PI controller is sufficient to fulfil the requirements of system stability and suitable dynamic response. Controller constants are usually identified using traditional methods such as the root locus method or frequency response methods, namely polar plots, Bode plots, etc. While these conventional methods of control system design can be directly abstracted from the aforementioned time/frequency domain analysis, in most situations the design methodology proceeds on a trial-and-error basis to meet certain performance specifications of the system. One of the difficulties of these methods is that a variety of control constants satisfy the performance requirements of the system. Then, the experience of the designer is applied to fine-tune the control constants.

However, the main hindrance in the preceding analysis is the requirement that a plant model is used in conjunction with a controller. The plant can be represented either as a transfer function model or as a state-space model. A few plants can be represented by mathematical or analytical equations and can be incorporated into a control system design process. However, the majority of plants cannot be easily represented either in transfer function form or in state-space form. Even if the models are available, they are partially true and are developed based on several assumptions. For example, designing a closed loop control for washing clothes in a washing machine cannot be represented by simple equations. Many inputs are very vague in nature. Other examples include complex chemical processes, waste water treatment plants, biomedical applications, scheduling problems and handwritten language detection. These systems are beyond the reach of conventional modelling and analysis. Thus, conventional design methods can be applied to those plants where a standard model exists or the model can be easily extracted. In other cases, it is the experience and intuition of the designer that pave the way for a robust control system design.

3.3 MAJOR COMPONENTS OF A FUZZY LOGIC SYSTEM

The "controller" in Figure 3.1 is now replaced with a fuzzy logic controller. For a fuzzy logic controller, there is no necessity for a plant model. The plant can be single input single output (SISO) or multi input multi output (MIMO). Human experience and expertise derived from association with the plant lead to the design of a fuzzy logic controller. Unlike a traditional controller, a fuzzy logic controller contains three major operations: fuzzification, inference engine and defuzzification, as shown in Figure 3.2. Each block is explained in the following sections.

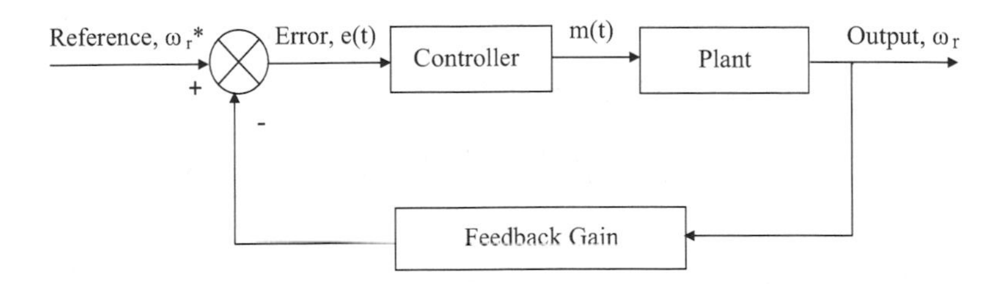

FIGURE 3.1 Closed loop system.

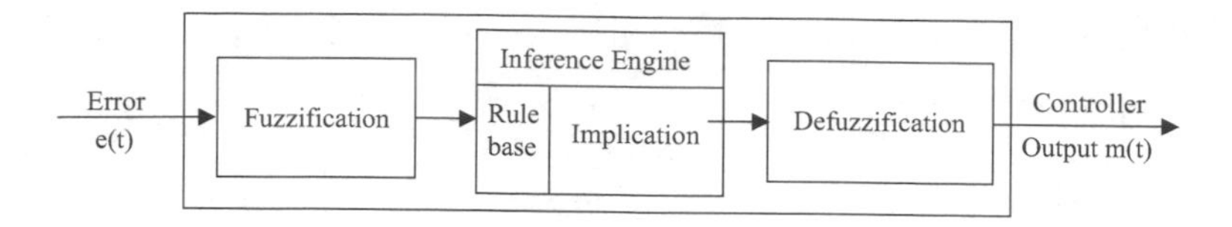

FIGURE 3.2 Fuzzy logic controller.

3.3.1 Fuzzification

In the real world, variables are measured and represented by numerical values, such as a speed error of 2 rpm or a temperature difference of 6.3°C. In a fuzzy logic system, numerical values have no significance, so these values cannot be used. Hence, in a fuzzy logic system, crisp inputs are converted into fuzzy variables for further processing. The crisp values of the input variable entering a fuzzy logic system are called fuzzy singletons. The process of converting a fuzzy singleton into a membership grade in one or many fuzzy sets is termed "fuzzification". For illustrations, consider Figure 3.3.

In Figure 3.3, the motor speed is described by three fuzzy sets labelled LOW, MED (medium) and HIGH. At the instant of measurement, when the fuzzy logic system is actually put into operation, the motor speed is, say 900 rpm. As can be seen from Figure 3.3, this crisp input variable has a membership grade of 0.4 in the set LOW and 0.8 in the set MED.

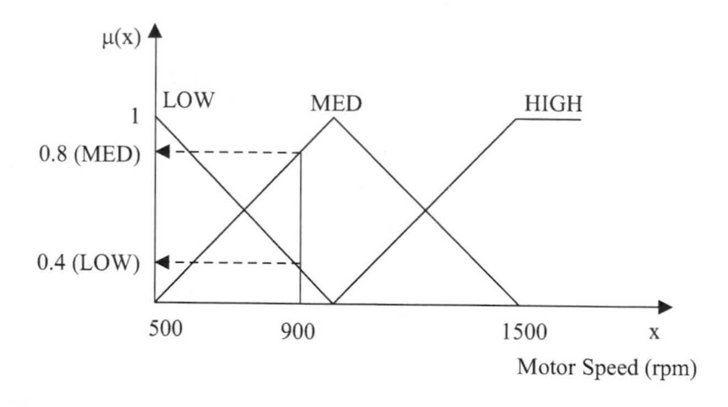

FIGURE 3.3 Fuzzification.

In other words, 900 rpm ⇒ 0.8 (MED)/0.4 (LOW).

These two membership grades will be used in the fuzzy logic processing system and no significance will be attached to the crisp value of 900 rpm hereafter. Thus, the fuzzification of 900 rpm results in membership grades in the respective fuzzy sets.

3.3.2 Inference Engine

The inference engine mainly consists of two sub-blocks, namely fuzzy rule base and fuzzy implication, as shown in Figure 3.2. The fuzzified inputs are fed to the inference engine and the rule base is then applied. The output fuzzy sets are subsequently identified using the implication method.

3.3.2.1 Rule Base or Fuzzy Propositions

At the heart of the fuzzy logic system is a rule base. A rule base consists of a set of fuzzy propositions and is derived from knowledge about the system. A fuzzy proposition or a statement establishes a relationship between different input fuzzy sets and output fuzzy sets.

A fuzzy model consists of a series of conditional and unconditional fuzzy propositions. A proposition or statement establishes a relationship between a value in the underlying domain and a fuzzy space.

(a) Unconditional Fuzzy Propositions: An unconditional fuzzy proposition is one that is not qualified by an *if* statement. The proposition has the general form:

x is Y

where x is a scalar from the domain and Y is a linguistic variable. Unconditional statements are always applied within the model and, depending on how they are applied, serve either to restrict the output space or to define a default solution space (if none of the conditional rules executes).

(b) Conditional Fuzzy Propositions: A conditional fuzzy proposition is one that is qualified by an *if* statement. The proposition has the general form:

If w is Z then x is Y

where w and x are model scalar values and Z and Y are linguistic variables. The proposition following the *if* term is the *antecedent* or *predicate*. The propositions following the *then* term is the *consequent*. A typical example of a fuzzy proposition is

If the height is TALL, then the weight is HEAVY

Here, TALL and HEAVY are input and output fuzzy sets describing the height and weight of an individual, respectively. This proposition states that if a person is TALL, he/she is likely to be HEAVY. This is called a conditional fuzzy proposition since it is qualified by an *if* statement. The foregoing fuzzy proposition is meant for an SISO system. In the case of multiple input single output (MISO) systems, the general structure of a fuzzy proposition is

$$\text{If } (\text{a is B}) \text{ AND } (\text{c is D}) \text{ AND } \ldots\ldots\ldots\ldots (\text{p is Q}), \text{ then x is Y}$$

where AND stands for the fuzzy AND operation. It can also be the OR operator depending on the type of problem.

(c) Rule Base Development: Fuzzy propositions are generally derived from a designer's experience and learning ability. Thus, the rule base design or development of fuzzy propositions for a system requires a thorough knowledge of and expertise in system characteristics.

Illustration 1

In order to illustrate this point, consider a simple example: Assume that you are going to a hotel for dinner with your friends.

At the end of the meal, you are planning to give a tip to the food server. The amount of the tip is influenced by many factors, but can be confined to two primary factors, namely the quality of the food and the service rendered. Thus, if the service is excellent and the quality of the food is excellent, the tip will also be excellent. On the other hand, if the quality of the food and the service is bad, naturally the tip will be low.

Thus, there are two inputs influencing the output. Consider three fuzzy sets describing the quality of the food. These fuzzy sets are labelled POOR, GOOD and EXCELLENT. Similarly, the other input variable, service, is also represented by three fuzzy sets with the semantic labels AVERAGE, GOOD and VERY GOOD. The tip amount is characterised by three fuzzy sets LOW, MED (medium) and HIGH. The fuzzy representation of the input and output variables is shown in Figure 3.4. Note that there are three fuzzy sets for each input variable and, therefore, there will be nine propositions.

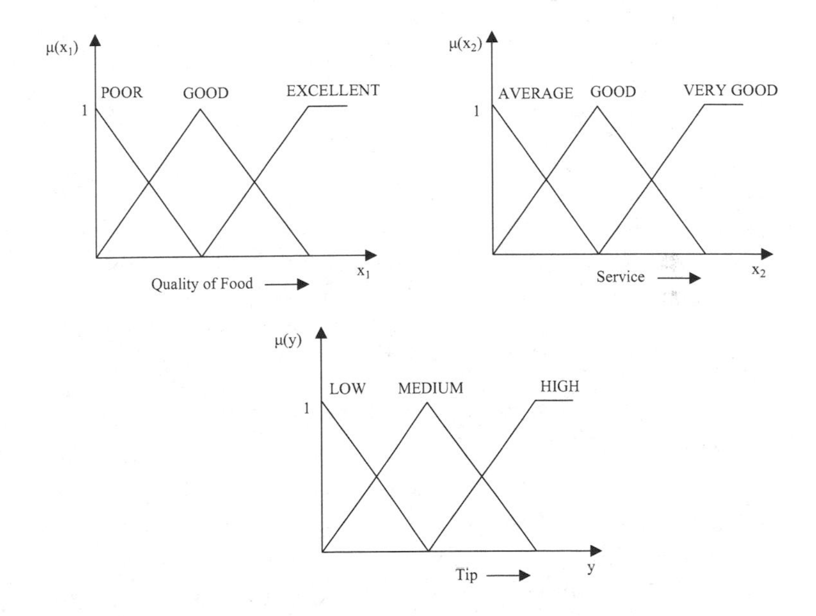

FIGURE 3.4 Fuzzy representation of input and output variables.

Thus, in general, if m is the number of fuzzy sets representing one input variable and n is the number of fuzzy sets for the second input variable for a two input-single output system, there will be a maximum of $m \times n$ propositions. However, not all the propositions need be valid and depend on the type of problem at hand. That is, the consequence of a certain proposition will not exist. Thus, certain propositions will be invalid as far as the system behaviour is concerned.

Certain valid propositions for the foregoing example are

If quality of food is EXCELLENT AND service is VERY GOOD, tip is HIGH.

If quality of food is EXCELLENT AND service is GOOD, tip is HIGH.

If quality of food is EXCELLENT AND service is AVERAGE, tip is MED.

If quality of food is GOOD AND service is VERY GOOD, tip is HIGH.

If quality of food is GOOD AND service is GOOD, tip is MED.

If quality of food is GOOD AND service is AVERAGE, tip is LOW.

If quality of food is POOR AND service is VERY GOOD, tip is MED.

If quality of food is POOR AND service is GOOD, tip is LOW.

If quality of food is POOR AND service is AVERAGE, tip is LOW.

It may be noted that the consequent fuzzy sets in these propositions may vary from individual to individual and also depend on the influence of each input variable on the individual. Thus, the rule base is time-variant, adaptive and robust in structure. When

TABLE 3.1 Fuzzy Associative Memory

	Quality		
Service	**POOR**	**GOOD**	**EXCELLENT**
AVERAGE	LOW	LOW	MED
GOOD	LOW	MED	HIGH
VERY GOOD	MED	HIGH	HIGH

the number of fuzzy sets describing input variables increases, the number of propositions also increases. Hence, it becomes a tiresome job of writing all the propositions sequentially. In order to avoid this difficulty, a matrix-like representation is generally followed and is known as fuzzy associate memory (FAM). The FAM for the preceding example is shown in Table 3.1.

Illustration 2

Consider another example, where a steady state speed estimation of a variable voltage direct current (dc) motor is carried out using the fuzzy logic system. A schematic of a variable armature voltage-variable load dc motor drive system is shown in Figure 1.1.

In Figure 1.1, the armature voltage, V_a, is varied to obtain a variable speed operation. The effect of the variation of V_a and the load current on the speed-current characteristics of the motor is depicted in Figure 1.1(b). Here, V_r represents the rated voltage. From the characteristics, it is evident that the motor speed, ω_r, is a function of V_a and the load current, I_a. Mathematically, we can write $\omega_r = f(V_a, I_a)$. This relationship is non-linear and mathematical simplification is difficult. A look-up table is one of the feasible ways out. However, it requires more memory and the resolution of the output depends on the amount of data stored. Fuzzy logic-based estimation is now developed with V_a and I_a as inputs and ω_r as output. The fuzzy sets are shown in Figure 3.5. In Figure 3.5, V_r is the rated voltage, I_r is the rated current and $\omega_{r(NL)}$ is the motor speed on no-load. A detailed examination of the speed-current characteristics of the motor shown in Figure 1.1(b) enables the rule base FAM as in Table 3.2.

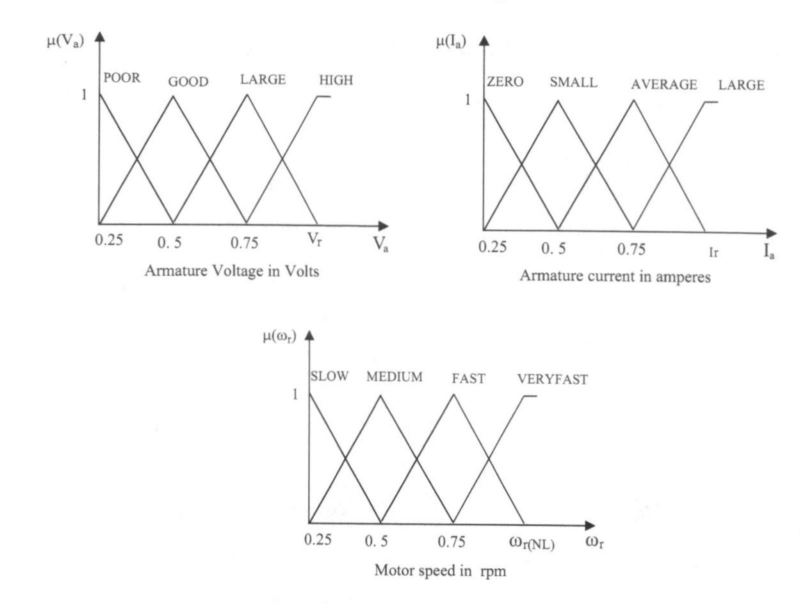

FIGURE 3.5 Fuzzy set for motor variables V_a, I_a and ω_r.

TABLE 3.2 FAM for DC Motor Steady State Speed Estimation

	Ia			
Va	**ZERO**	**SMALL**	**AVERAGE**	**LARGE**
LOW	FAST	MEDIUM	SLOW	SLOW
MEDIUM	FAST	FAST	MEDIUM	SLOW
LARGE	VERY FAST	FAST	FAST	MEDIUM
HIGH	VERY FAST	VERY FAST	FAST	FAST

3.3.2.2 Min-Max Method of Implication

This implication method is used with MIMO systems where the input and output fuzzy sets are more than one. Thus, the rule base has several propositions. It is important to note that this implication method must be rigorously followed by the defuzzification procedure. The min-max compositional operation derives its name from the procedure followed in the method. The consequent fuzzy region is restricted to the minimum of the predicate truth while selecting the output fuzzy set; the output fuzzy region is updated by taking the maximum of these minimised fuzzy sets during construction of the output fuzzy space. These steps result in reducing the strength (that is, the effective height) of the fuzzy set output to equal the maximum truth of the predicate. When all the propositions have been evaluated, the output contains a region or space in the output fuzzy set which reflects the contribution from each proposition.

This method consists of two steps:

(i) Identifying the output fuzzy region and then

(ii) Configuring the output fuzzy region.

The two steps are illustrated as follows:

Consider the steady state speed estimation of a variable voltage variable load dc motor drive system. The two inputs are the motor terminal voltage and the current and the output is the motor speed. The three variables are redrawn in the fuzzy context and are given in Figure 3.6.

There are four fuzzy sets for each input variable, accounting for a maximum of 16 propositions. At the instant of measurement, let the current be 3.4 A and the voltage be 170 V. The following propositions are activated alone. The numerical values within parentheses indicate the membership grades for these two fuzzy singletons, which are shown in Figure 3.6(a).

- If voltage is MED (0.3) AND current is AVERAGE (0.5), speed is MEDIUM.
- If voltage is LARGE (0.8) AND current is SMALL (0.7), speed is FAST.

When the first proposition is fired, the truth of the predicate is taken as the minimum of the membership grades of the two input fuzzy sets, i.e. $0.3 \cap 0.5 = 0.3$. Thus, the consequent fuzzy set,

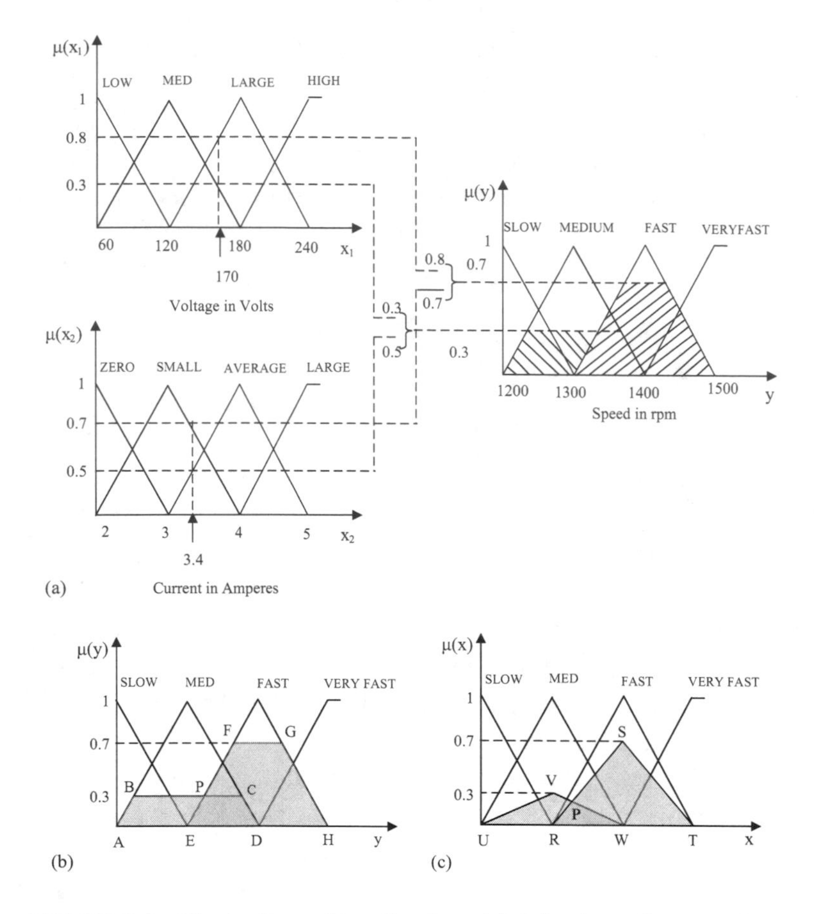

FIGURE 3.6 Illustration of implication. (a) Selection of output fuzzy sets. (b) Correlation minimum.

MEDIUM, is selected at 0.3. In a similar way, when the next rule is activated, 0.8 and 0.7 are ANDed giving the result of 0.7. Thus, the output fuzzy set FAST is selected at 0.7. Therefore, depending upon the value of the fuzzy singletons, respective fuzzy rules are fired. The consequent fuzzy sets are identified at the truth of the predicate. Once the output fuzzy sets are identified, configuring or shaping the output fuzzy region is the next step. This is done using either of the following methods:

(a) *Correlation minimum:* Here, the output fuzzy sets (MEDIUM and FAST) are truncated at the respective truth levels, i.e. MEDIUM is chopped at 0.3 and FAST is truncated at 0.7, resulting in trapezoidal fuzzy sets, namely ABCD and EFGH. This is shown in Figure 3.6(b). The aggregate output fuzzy region is obtained by the union of ABCD and EFGH. This results in a polygon with edges ABPFGH. Thus, in the present case, for a current of 3.4 A and 170 V, the respective output fuzzy region is configured.

(b) *Correlation product:* In this process, the original shapes of the selected fuzzy sets are retained at the truth of the predicate. In the current example, the output fuzzy set MEDIUM is scaled down to another triangular fuzzy set UVW with its peak value as 0.3, i.e. truth of the predicate. Similarly, the fuzzy set FAST is reshaped to the triangular fuzzy set RST of maximum value 0.7, as shown in Figure 3.6(c). The aggregate output solution space is obtained as in the previous case. The resultant solution space is a polygon with edges UVPST. A comparison between the correlation minimum method and the correlation product method is not wise. Depending on the application, the user may choose either one. However, some theoreticians argue that while doing defuzzification, the correlation minimum method gives a reduced computational burden since a truncated output fuzzy set can be easily perceived and used for computation whereas the output fuzzy set in the correlation product method needs to be "reconstructed" for defuzzification. A few researchers believe that the correlation product method "truly reflects" the output fuzzy region since it retains the original shape.

It may be noted that while extracting truth from the predicate, we go for ANDing (i.e. min) the membership grades and while finding the aggregate output fuzzy region, we perform the OR (i.e. max) operation. Hence, the name min-max method is derived.

3.3.3 Defuzzification Methods

After fuzzy implication, the output fuzzy region is constructed. Extraction of the numerical value corresponding to the output from the output fuzzy region is called defuzzification. There are different defuzzification schemes. A few important and commonly employed defuzzification methods are discussed as follows:

(a) Centre of area/gravity method

(b) Centre of sums method

(c) Height method

(d) Centre of largest area method

(e) First of maxima method

(f) Last of maxima method

(g) Middle of maxima method

Prior to a discussion of each method, certain common symbols and notations will be given for ease of understanding. After fuzzy implication, the output fuzzy sets are either truncated (clipped) or scaled down. In general, it is assumed that the fuzzy sets are clipped.

Referring to Figure 3.7(a–c) and, let us use the following notations:

CL: Clipped fuzzy set; this is marked as CL(1), CL(2), CL(m),…, CL(n).

m: Number of propositions activated at a given instant; m = 1 to n, which is also equal to the number of clipped fuzzy sets.

U: Output solution space, comprising all clipped fuzzy sets; this is divided into l discrete values.

u_i: Discrete value within U, where i = 1 to l.

$\mu(u_i)$: Membership grade of u_i in the original (unclipped) fuzzy set.

$\mu_{CL}(u_i)$: Membership grade of u_i in the clipped fuzzy set.

$\mu_{CL(m)}(u_i)$: Membership grade of u_i in the m-th clipped fuzzy set.

u*: Defuzzified output.

In Figure 3.7(a), a valid proposition truncates the output fuzzy set HIGH at 0.2. Thus, ABCD is the clipped fuzzy set to be used for configuring the output fuzzy set. Any output variable u_i within the domain of HIGH has two membership values now, one in the original fuzzy set (i.e. in HIGH), which is marked as $\mu(u_i)$, and the other in the clipped fuzzy set indicated as $\mu_{CL}(u_i)$, which is 0.2 in this case.

Consider a case where three propositions are fired and let the truth of the predicate in each one be 0.75, 0.5 and 0.2, as shown in Figure 3.7(b). Thus, Figure 3.7(b) shows three clipped fuzzy sets, namely ABCD, EFGH and DJKL, all trapezoidal in shape. These are marked as CL(1), CL(2) and CL(3), respectively. The output fuzzy region encompasses all three clipped fuzzy sets and is marked as ABPFGJKL. Hence, all variables from A to L form the solution space U. The defuzzified value, u*, falls within U.

3.3.3.1 Centre of Area Method or Centre of Gravity Method

This method is commonly employed and is observed to be very effective. Consider the solution space U in Figure 3.7(b). Assume that U can be divided into l discrete values, say $u_1, u_1, u_1 \ldots u_1 \ldots u_1$, where i = 1 to l. Each u_i has a membership value of $\mu_{CL}(u_i)$ in one alone or in many clipped fuzzy sets. In this method, the membership of a variable u_i in a clipped fuzzy set is taken as such, if the variable belongs to only one clipped fuzzy set. If the variable belongs to many clipped fuzzy sets (i.e. in an overlapping region of clipped fuzzy sets), the maximum membership of the variable among all the clipped fuzzy sets is considered its membership value for defuzzification. Thus, the membership grades of u_i are ORed to perform defuzzification in the overlapped regions.

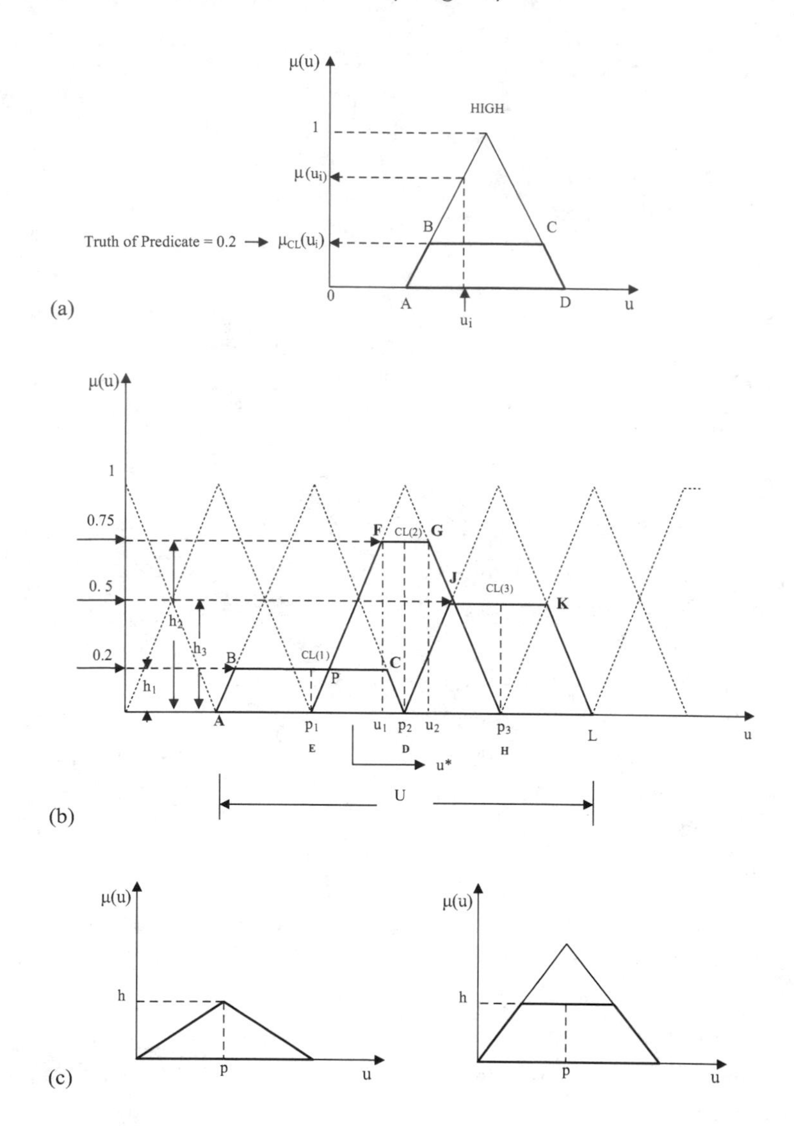

FIGURE 3.7 Definition of symbols (a) $\mu(u_i)$, $\mu CL(u_i)$. (b) Clipped sets CL(1), CL(2) and CL(3). (c) Height and peak values.

Hence, in the centre of gravity method, the defuzzified output u* is given by Equation 3.4.

$$u^* = \frac{\sum_{i=1}^{l} u_i \bigcup_{m=1}^{n} \mu_{CL(m)}(u_i)}{\sum_{i=1}^{l} \bigcup_{m=1}^{n} \mu_{CL(m)}(u_i)} \tag{3.4}$$

There is an alternative way to express u*. Consider a fuzzy set CL, which is the ORed version of all clipped fuzzy sets.

Thus, $CL = CL(1) \cup CL(2) \cup \ldots CL(K) \cup \ldots \cup \ldots CL(n)$

Then

$$u^* = \frac{\sum_{i=1}^{l} u_i \mu_{CL}(u_i)}{\sum_{i=1}^{l} \mu_{CL}(u_i)} \tag{3.5}$$

3.3.3.2 Centre of Sums Method

In this method, the identity of each fuzzy set is retained, i.e. the membership grades of the variable are not ORed. Thus, overlapping portions of the adjacent clipped fuzzy set appear more than once. Accordingly, the expression for u* is given in Equation 3.6.

$$u^* = \frac{\sum_{i=1}^{l} u_i \sum_{m=1}^{n} \mu_{CL(m)}(u_i)}{\sum_{i=1}^{l} \sum_{m=1}^{n} \mu_{CL(m)}(u_i)} \tag{3.6}$$

3.3.3.3 Height Method

The height, h, and the peak value, p, of fuzzy sets are used in this method. These parameters are defined as follows:

The height, h, of a fuzzy set is defined as the maximum membership grade of the fuzzy set. In a clipped fuzzy set, this value is equal to the truth of the predicate. The peak value, p, of a fuzzy set is the variable within the domain of the fuzzy set having the largest membership grade. In the case of a triangular fuzzy set, this value is unique whereas with a trapezoidal set, this is an interval. Instead of taking the interval, the midpoint of this interval is taken as the peak value. This is important to mention because clipped fuzzy sets are trapezoidal in shape. This is illustrated in Figure 3.7(c).

Now consider Figure 3.7(b), where three propositions generate three clipped fuzzy sets. Each set is characterised by height and peak values.

In the height method, u* is given by Equation 3.7:

$$u^* = \frac{p_1 h_1 + p_2 h_2 + p_3 h_3}{h_1 + h_2 + h_3} \tag{3.7}$$

In general, defuzzified value has the following form:

$$u^* = \frac{\sum_{m=1}^{n} p_m h_m}{\sum_{m=1}^{n} h_m} \tag{3.8}$$

The main advantage of this method is its computational simplicity. The computational burden is significant with the centre of gravity and the centre of sums methods.

3.3.3.4 Centre of Largest Area Method

In this defuzzification procedure, the clipped fuzzy set with the largest area is selected. For example, in Figure 3.7(b), CL(2) has the maximum area. Now, u* is computed by applying the centre of gravity method to this clipped fuzzy set alone.

3.3.3.5 First of Maxima Method

In this method, the clipped fuzzy set having the highest membership grade is first identified. Then, the smallest value of the variable within the domain of this clipped fuzzy set with the highest membership grade is taken as u*. In Figure 3.7(b), CL(2) is the clipped fuzzy set with the maximum membership value and u_1 is the smallest variable with this membership. Thus, u_1 is taken as u*. Let u_x denote the domain of the clipped fuzzy set with the largest membership grade. Mathematically, we can write this procedure as

Let $K = \max_{U} \{\mu_{CL}(u)\}$ be the largest membership grade in solution space U.

Then $u^* = \min_{u_x} \{u \mid \mu_{CL}(u) = K\}$.

This is u_1 as in Figure 3.7(b).

3.3.3.6 Last of Maxima Method

This is similar to the previous method. The clipped fuzzy set with the largest membership grade is identified and the highest value of the variable in its domain with the maximum membership grade is taken as u*. In Figure 3.7(b), u_2 is the defuzzified value.

Mathematically

$$u^* = \max_{u_x} \{u \mid \mu_{CL}(u) = K\} \tag{3.9}$$

3.3.3.7 Middle of Maxima Method

Here, the average value of u_1 and u_2 is computed as u*.

$$u^* = \frac{\min_{u_x} \{u \mid \mu_{CL}(u) = K\} + \max_{u_x} \{u \mid \mu_{CL}(u) = K\}}{2} \tag{3.10}$$

3.4 COMPARISON AND EVALUATION OF DEFFUZIFICATION METHODS

In this section, we suggest some criteria which an "ideal" defuzzification method should satisfy. It must be stated in advance that none of our defuzzification methods satisfies all of the following criteria, i.e. one has to weigh these criteria for the particular application to make the right choice of the defuzzification method.

3.4.1 Continuity

A small change in the input of the fuzzy system should not result in a large change in the output. For example, in the case of a two input-single output fuzzy system, when two subsequent inputs (x_1, y_1) and (x_2, y_2) differ slightly, the corresponding output values, u_1^* and u_2^*, should also differ very slightly. Mathematically, we can write this criterion as

$$\text{For } |x_1 - x_2| \to 0 \text{ and } |y_1 - y_2| \to 0.$$

$$\text{Then } |u_1^* - u_2^*| \to 0.$$

3.4.2 Unambiguity

While proceeding with defuzzification, there should not be any ambiguity hindering the process. Consider the application of the centre of largest area method to the two clipped fuzzy sets CL(2) and CL(3), shown in Figure 3.7(b). Here, both CL(2) and CL(3) have the same area, thus the centre of largest area method cannot be proceeded further. A similar problem arises with the first of maxima, the last of maxima and the middle of maxima methods, when applied to Figure 3.7(b), since the two clipped fuzzy sets have the same maximum membership grade. Thus, it can be concluded that the centre of area, the centre of sums and the height methods possess unambiguity.

3.4.3 Plausibility

When clipped fuzzy sets are identified, the solution space U extends from the leftmost clipped fuzzy set to the rightmost clipped fuzzy set. The defuzzified value u* emerges within U. A defuzzification procedure is said to be plausible if the following conditions are satisfied:

- u* lies in the middle of U and
- u* has the largest membership grade in the clipped fuzzy set.

It may be noted that the centre of gravity and the centre of sums methods satisfy this criterion to a large extent, while other methods need not be plausible.

3.4.4 Computational Complexity

This is important in the case of online process control. A review of various methods shows that the height method is the fastest method followed by first of maxima, last of maxima and middle of maxima. The centre of largest area method is slow, but the centre of sums method is even slower. The computational burden is highest with the centre of area method. However, the process time constant can be taken as the base time while determining the computational time requirement of each defuzzification scheme.

3.4.5 Weight Counting

This property enables the contribution of each clipped fuzzy set to the defuzzified output, u*. Thus, in the case of the centre of area, the centre of sums and the height methods, each clipped fuzzy set is involved in the calculation of u*. The other methods, however, do not exhibit this property.

3.5 CONCLUSION

In this chapter, a detailed discussion on the functioning of a fuzzy logic system has been presented. Each block in a fuzzy logic controller has been elaborated through examples and illustrations.

QUESTIONS

1. Sketch a typical closed loop system for a process control and explain the difficulties in tuning the controller.
2. Using a sketch, explain the major components of a fuzzy logic system.
3. Illustrate the fuzzification process.
4. Write a conditional fuzzy statement.
5. Write an unconditional fuzzy proposition.
6. Demonstrate conditional and unconditional fuzzy propositions with suitable examples.
7. Demonstrate rule base development using an appropriate example.
8. Develop a rule base for the speed estimation of a dc motor drive.
9. Develop a fuzzy logic–based efficiency estimation for a fixed voltage variable load induction motor drive.
10. How will you develop the rule base for the operation of an air-conditioner? Explain.
11. What is FAM?
12. What is the maximum number of rules for a two input-single output system with each input described by three fuzzy sets?
13. Elaborate on the min-max method of implication.
14. Illustrate the correlation minimum and correlation product methods.
15. What is the advantage of the correlation minimum method?

16. Define the process of defuzzification.
17. Using figures and symbols, explain any five methods of defuzzification.
18. Compare and evaluate different defuzzification schemes.
19. What is the major disadvantage of an increased number of fuzzy sets?
20. Name the most favoured defuzzification scheme.
21. Define the "height" of a fuzzy set.
22. What is the major drawback of the last of maxima method?
23. What is the meaning of the term "weight counting" in connection with defuzzification?
24. Which defuzzification is computationally most difficult?
25. What is the major drawback of the centre of area method?
26. Prove with necessary derivations/expressions that a fuzzy controller is inherently non-linear.
27. In the closed loop speed control of a drive, the output variable is speed and the inputs are voltage and current. Speed is described by several fuzzy sets and two adjacent fuzzy sets for the speed are MED and HIGH: these sets are symmetric and have the domains (1000–1200 rpm) and (1100–1300 rpm), respectively. Compute the speed of a motor if the following fuzzy rules are activated:

    If voltage is LOW (0.72) AND current is VERY SMALL (0.26), then speed is MED.

    If voltage is AVG (0.34) AND current is SMALL (0.66), then speed is HIGH.

    Employ a correlation product for implication and the height method for defuzzification.

28. In a temperature control scheme, the inputs are current and voltage and the output is temperature. The temperature is described by several fuzzy sets. Two fuzzy sets, WARM and HOT, are adjacent and have the domains (100°C–200°C) and (150°C–250°C), respectively. At a typical operating point, the following rules are activated:

    If voltage is LOW (0.72) AND current is VERY SMALL (0.26), then temperature is WARM.

    If voltage is AVG (0.34) AND current is SMALL (0.66), then temperature is WARM.

    Employing a correlation product for implication and the first of maxima method for defuzzification, find the temperature.

CHAPTER 4

Design Aspects of Fuzzy Systems and Fuzzy Logic Applications

4.1 INTRODUCTION

This chapter deliberates on various methods of design of a fuzzy logic–based system. These include membership function construction, rule base design, etc. A few fuzzy logic applications are also presented.

4.2 SUGGESTIONS ON FUZZY SET DESIGN

For a beginner, it is preferable to follow certain guidelines while designing a fuzzy logic system:

- Use uniform fuzzy sets, i.e. employ similar types of membership functions to describe a single variable.

- Use fuzzy sets that can be expressed in simple mathematical terms. Triangular fuzzy sets are preferable in this regard.
- Employ symmetrical fuzzy sets.

  A fuzzy set is said to be symmetric if its left width is equal to its right width. This property enables the defuzzified output, u*, to fall in the middle of the clipped fuzzy set, thereby satisfying plausibility.

- The fuzzy sets should overlap as discussed in Section 2.4.5. If fuzzy sets do not overlap, there is an uncovered domain of variability, which does not occur in either fuzzy set. Thus, when a fuzzy singleton occurs in this domain, no proposition will be activated, resulting in no change in control action. This can cause a poor response or even instability.
- *Quantum of overlap*: Ideally 50% overlap is generally advised. However, if this does not give satisfactory results and if the fuzzy sets are non-symmetric, the following rule is applied:

  The right width of a fuzzy set should equal the left width of the next adjacent fuzzy set and both should equal the interval between the peak values of the two fuzzy sets.

- *Number of fuzzy sets*: This is purely problem dependent. Less fuzzy sets require less memory for storing fuzzy associative memory (FAM). However, this can reduce the quality of the system's response. An increase in the number of fuzzy sets increases the output precision, leading to a good dynamic response. However, an increase in the number of fuzzy sets will require more memory space and computing time. Thus, the number of fuzzy sets should be judiciously chosen.

4.3 EXTRACTING INFORMATION FROM KNOWLEDGE ENGINEER

In this section, we will discuss developing fuzzy sets by interacting with people who are exposed to the problem under consideration for a long time. Such individuals are called "experts" or "knowledge engineers" in this regard. A fuzzy set designer can ask an expert suitable questions to arrive at the following:

- Labels to be used

The expert can suggest appropriate semantic titles. Labels should be consumer-friendly and appropriate for the application. For automatic washing machines, it is preferable to put "mild wash", "hard wash", "normal wash", etc. This is similar to labelling "wool", "cotton", etc. on an iron.

- Domain of a fuzzy set

Once the universe of discourse of the variable concerned is known, it can be divided into various fuzzy sets. The domain of each fuzzy set can be obtained by posing questions such as

Does 20°C indicate "COLD" in an air-conditioning system?

Is 100 rpm suitable for a "HARD WASH" for a washing machine?

Once the approximate domain is located, an expert or experts can be consulted further to ascertain the membership grade of each variable. A single expert can be asked to allocate grades for each variable by posing questions such as

> "What would be the membership grade of 20°C in 'COLD'?"

or

> "What would be the possible membership grade of 50 rpm in 'NORMAL WASH'?"

Once the membership grades have been obtained, a scattered pattern can be viewed to find a suitable standard fuzzy set to represent it. Then, the membership grades are used to construct a proper fuzzy set by using the curve fitting method.

4.4 ADAPTIVE FUZZY CONTROL

A fuzzy logic system is adaptive to plants that exhibit parameter variation and non-linearity. However, if a plant's coefficients vary largely, a simple fuzzy control may not be effective. In such cases, it is possible to develop adaptive fuzzy control methods.

Generally, fuzzy sets in a fuzzy logic system have a fixed domain. It is possible to alter the domain of all fuzzy sets by employing a scaling factor. This is illustrated in Figure 4.1. Here, the input variable is multiplied by a factor, K, prior to fuzzification. As shown in Figure 4.1, for K=1, the universe of discourse is from −100 to +100, whereas it changes from −50 to +50 for K=0.5 and from −0.1 to +0.1 for K=0.1. The variation of K results in the classification of an input variable. For example, when the input variable value is 70 for K=1, the variable belongs to fuzzy sets PM to PB. The same variable belongs to PS and PM fuzzy sets with K=0.5, whereas it occurs in the domain of ZERO and PS for K = 0.1. Thus, altering the value of the scaling factor leads to activation of different fuzzy prepositions for the same input value. Thus, if the plant parameters vary largely, suitable variation can be adopted for K leading to adaptive fuzzy control.

4.5 FUZZY DECISION-MAKING

Decision-making is a common activity in people's daily life; therefore, it can be treated as one of the most fundamental activities.

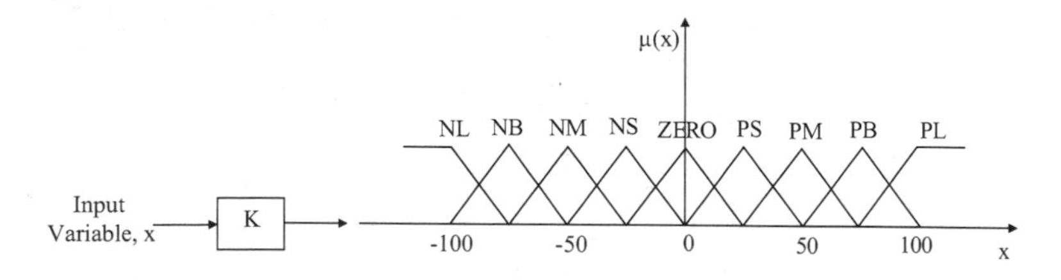

FIGURE 4.1 Adaptive fuzzy control.

Decision-making is an area that studies how decisions are actually made and how better decisions can be successfully made. A broad look at how a decision is made reveals that it is the best choice among various alternatives available.

Each option is associated with a certain quantum of desired results as well as constraints. The last two factors are generally weighted terms, weight being a variable with each alternative. Thus, the best choice or design is one that contains the maximum number of desired objectives and minimum number of constraints/disadvantages. Thus, decision-making can well be considered an optimisation problem.

Fuzzy logic concepts can be employed to arrive at a suitable decision by incorporating an individual's perception about alternative options and their constraints as simple fuzzy sets. Two major advantages of fuzzy decision-making are

- A simple, single fuzzy set – a freehand sketch in many cases – is sufficient to absorb vagueness connected with each input and its constraints.
- The procedure is systematic and leads to a near-optimal solution.

A decision-making process, as per the fuzzy logic perspective, includes the following components:

(1) A group of options O:

$$O = \{o_1, o_2, o_3, \ldots, o_i, \ldots, o_l\}$$

$$O = \{o_i\}, \quad \text{for } i = 1, 2, \ldots, l$$

(2) Set G describing the goals associated with each option:

$$G = \begin{bmatrix} g_1/o_1 & g_1/o_2 & \cdots & g_1/o_i & \cdots & g_1/o_l \\ g_2/o_1 & g_2/o_2 & \cdots & g_2/o_i & \cdots & g_2/o_l \\ \cdots & \cdots & \cdots & \cdots & \cdots & \cdots \\ g_j/o_1 & g_j/o_2 & \cdots & g_j/o_i & \cdots & g_j/o_l \\ \cdots & \cdots & \cdots & \cdots & \cdots & \cdots \\ g_m/o_1 & g_m/o_2 & \cdots & g_m/o_i & \cdots & g_m/o_l \end{bmatrix}$$

$$= \{g_j \mid o_i\}, \text{ where } j = 1 \text{ to } m$$

(3) Set C describing the constraints associated with each option:

$$G = \begin{bmatrix} c_1/o_1 & c_1/o_2 & \& & c_1/o_i & \& & c_1/o_l \\ c_2/o_1 & c_2/o_2 & \& & c_2/o_i & \& & c_2/o_l \\ \& & \& & \& & \& & \& & \& \\ c_k/o_1 & c_k/o_2 & \& & c_k/o_i & \& & c_k/o_l \\ \& & \& & \& & \& & \& & \& \\ c_n/o_1 & c_n/o_2 & \& & c_n/o_i & \& & c_n/o_l \end{bmatrix}$$

$$= \{c_k \mid o_i\}, \text{ where } k = 1 \text{ to } n$$

Now, the goals and constraints are to be developed as fuzzy sets. Generally, each goal/constraint requires only one fuzzy set. The fuzzy sets are to be delineated as per the priorities of an individual. Hence, it may be noted that fuzzy decision-making is subjective in nature.

The membership grade of each o_i in each goal/constraint is determined, i.e. $\mu(g_j/o_i)$ and $\mu(c_k/o_i)$ are determined.

Let

$$\underset{i=1\text{ to }l}{D}(o_i) = \min_{\substack{i=1\text{ to }l\\ j=1\text{ to }m\\ k=1\text{ to }n}} \left\{ \mu\left({}^{g_j}\!/_{o_i} \right), \mu\left({}^{c_k}\!/_{o_i} \right) \right\} \tag{4.1}$$

Then, the fuzzy decision is given by

$$D = \max_{i=1\text{ to }l} \left\{ D(o_i) \right\} \tag{4.2}$$

PROBLEM 1

A research institute wants to recruit a young, dynamic and talented scientific officer to assist a team comprising three experts to work in the field of high voltage engineering.

The institute is experiencing financial constraints. If the institute wishes to call only one person for interview, develop a fuzzy decision-making algorithm and give the solution for the present problem.

There are four applications and the details are as follows:

Name	Age	Qualification	Salary Demand
a1	25	MTech	Rs.12,500
a2	55	BTech	Rs.25,000
a3	32	PhD	Rs.17,000
a4	40	MTech	Rs.21,000

Solution

In Problem 1, the institute is looking for young, dynamic (basically age decides) and talented (better qualified) candidates. Therefore, it is appropriate to consider age and qualification as goals, namely g_1 and g_2.

The institute has financial impediments. Therefore, salary is constraint c_1. Thus, the problem has two goals and a single constraint. Let us build simple fuzzy sets g_1, g_2 and c_1. It may be noted that an increase in age is less preferred. Similar is the case with salary too. As qualification is higher, it is better. The three fuzzy sets are sketched as shown in Figure 4.2.

It is important to note that qualification cannot be expressed numerically. However, since a choice has to be made, this variable is assigned a suitable membership grade.

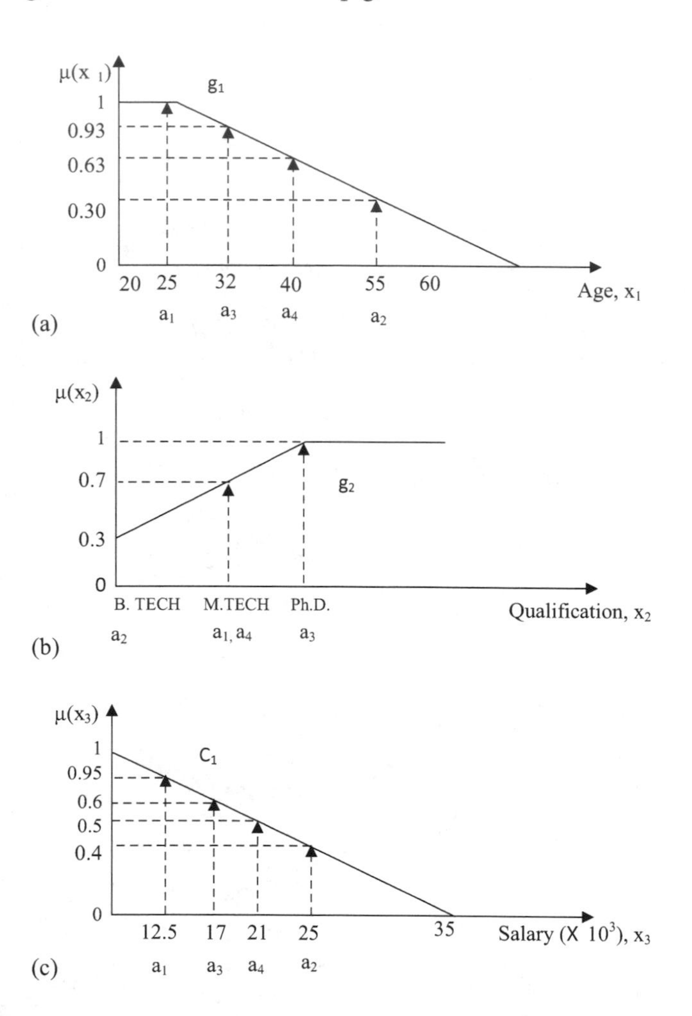

FIGURE 4.2 Fuzzy sets for Problem 1: (a) age, (b) qualification and (c) salary.

From Figure 4.2
we find

$\mu(g_1/a_1), \mu(g_1/a_2), \mu(g_1/a_3), \mu(g_1/a_4)$

$\mu(g_2/a_1), \mu(g_2/a_2), \mu(g_2/a_3), \mu(g_2/a_4)$

$\mu(C_1/a_1), \mu(C_1/a_2), \mu(C_1/a_3), \mu(C_1/a_4)$

$\mu(g_1/a_i) = [1/a_1, 0.3/a_2, 0.93/a_3, 0.63/a_4]$

$\mu(g_2/a_i) = [0.7/a_1, 0.3/a_2, 1.0/a_3, 0.7/a_4]$

$\mu(C_1/a_i) = [0.95/a_1, 0.4/a_2, 0.6/a_3, 0.5/a_4]$

Using Equation 4.1:

$$D(a_i) = [0.7/a_1, 0.3/a_2, 0.6/a_1, 0.5/a_1]$$

Now applying Equation 4.2, we get the fuzzy decision:

$$D = \text{Max}[D(a_i)] = 0.7/a_1$$

Thus, a_1 is the suitable candidate to be called for interview.

PROBLEM 2

After graduating, an engineering student is offered four different jobs. The characteristics of each job are as follows:

Job	Salary	Distance from Hometown	Nature of Appointment
J1	Rs 12,000	10 km	Permanent after one year
J2	Rs 20,000	200 km	Permanent after two years
J3	Rs 15,000	40 km	Permanent after two years
J4	Rs 27,000	300 km	Contract basis

Develop a fuzzy decision-making algorithm and find a suitable job.

Solution

Here, we select salary as a goal, i.e. g_1, and distance from hometown and nature of appointment are the constraints, i.e. c_1 and c_2. Fuzzy sets are sketched as shown in Figure 4.3.

From Figure 4.3, we find

$$\mu(g_1/J_1),\ \mu(g_1/J_2),\ \mu(g_1/J_3),\ \mu(g_1/J_4)$$

$$\mu(C_1/J_1),\ \mu(C_1/J_2),\ \mu(C_1/J_3),\ \mu(C_1/J_4)$$

$$\mu(C_2/J_1),\ \mu(C_2/J_2),\ \mu(C_2/J_3),\ \mu(C_2/J_4)$$

$$\mu(g_1/J_i) = [0.6/J_1,\ 0.8/J_2,\ 0.7/J_3,\ 0.95/J_4]$$

$$\mu(C_1/J_i) = [1/J_1,\ 0.4/J_2,\ 0.90/J_3,\ 0.1/J_4]$$

$$\mu(C_2/J_i) = [0.9/J_1,\ 0.8/J_2,\ 0.8/J_3,\ 0/J_4]$$

Using Equation 4.1:

$$D(j_i) = \left[{}^{0.1}\!/_{J_1}, {}^{0.4}\!/_{J_2}, {}^{0.8}\!/_{J_3}, {}^{0}\!/_{J_4}\right]$$

Now applying Equation 4.2, we get the fuzzy decision:

$$D = \max_{i=1 \text{ to } 4}(D(J_i)) = {}^{0.8}\!/_{J_3}$$

Hence, job J_3 is selected.

4.6 NEURO-FUZZY SYSTEMS

Similar to fuzzy logic, a neural network does not require a "model" of a plant to be controlled. The major advantage of fuzzy logic is that fuzzy rules in natural terms can be easily derived from an expert. The chief merits of a neural network is its learning ability and that it can approximate non-linear functions with better accuracy. Therefore, it is of great interest to see the result when

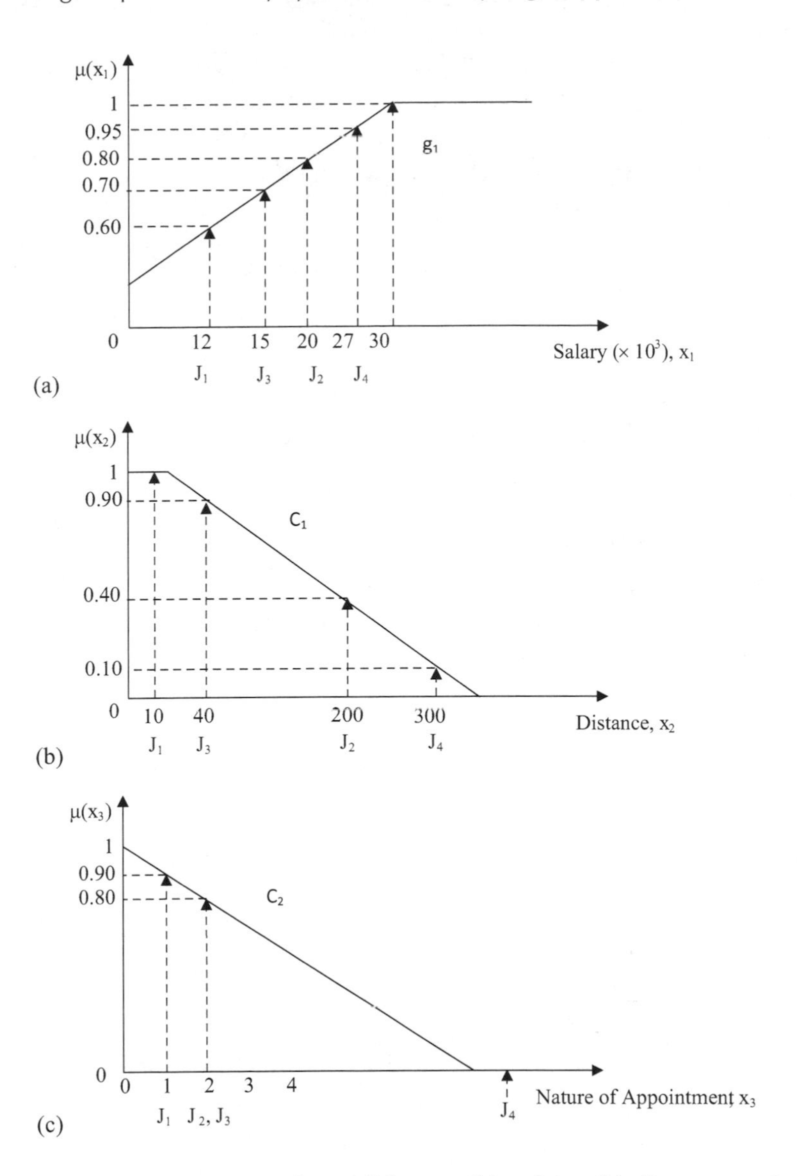

FIGURE 4.3 Fuzzy sets for Problem 2: (a) salary, (b) distance and (c) nature of appointment.

two "model-free" methodologies are combined. The resulting configuration is called a neuro-fuzzy system and it possesses the advantages of both fuzzy systems and neural networks. A brief description of a neural network is given prior to a description of a neuro-fuzzy system.

The basic concept of an artificial neural network is derived from a biological neuron. A simplified structure of a biological neuron is given in Figure 4.4(a).

The basic component of the brain is a neuron. A neuron consists of a cell body, which encompasses the nucleus. The interior of the neuron is filled with liquid called protoplasm. Nerve fibres emanate from the periphery of the cell body and split into branches and sub-branches called dendrites. These dendrites receive signals from other neurons. A single, long fibre from the neuron is called the axon, which also splits into roots and sub-roots. The axon carries information signals from one neuron to other neurons. The fibre branches of the axon are connected to dendrites of adjacent neurons. However, these two are not directly connected: there is a gap between the strands of the axon and the dendrites. This gap is called a synapse or a synaptic function.

A neuron's activity is a combined electrochemical process. Under idle conditions, the protoplasm is negatively charged against the surrounding neural fluid, which contains a substantial number of positive sodium ions. This causes a potential difference of about 70 mV between the interior and the exterior of a cell body. This electrical voltage is sustained during a neuron's period of inactivity due to the neuron body membrane, which is impenetrable to positively charged sodium ions. When a signal comes from the dendrites, this potential is disturbed. If the incoming signal is capable of lowering the potential to approximately 60 mV, the cell body membrane loses its strength, causing positive sodium ions to enter the neuron body. This sudden inrush causes a release of charges through the axon. When these charges reach the synaptic gap, the conductivity of the synaptic gap is varied by the chemical process, thereby allowing the charges to reach the

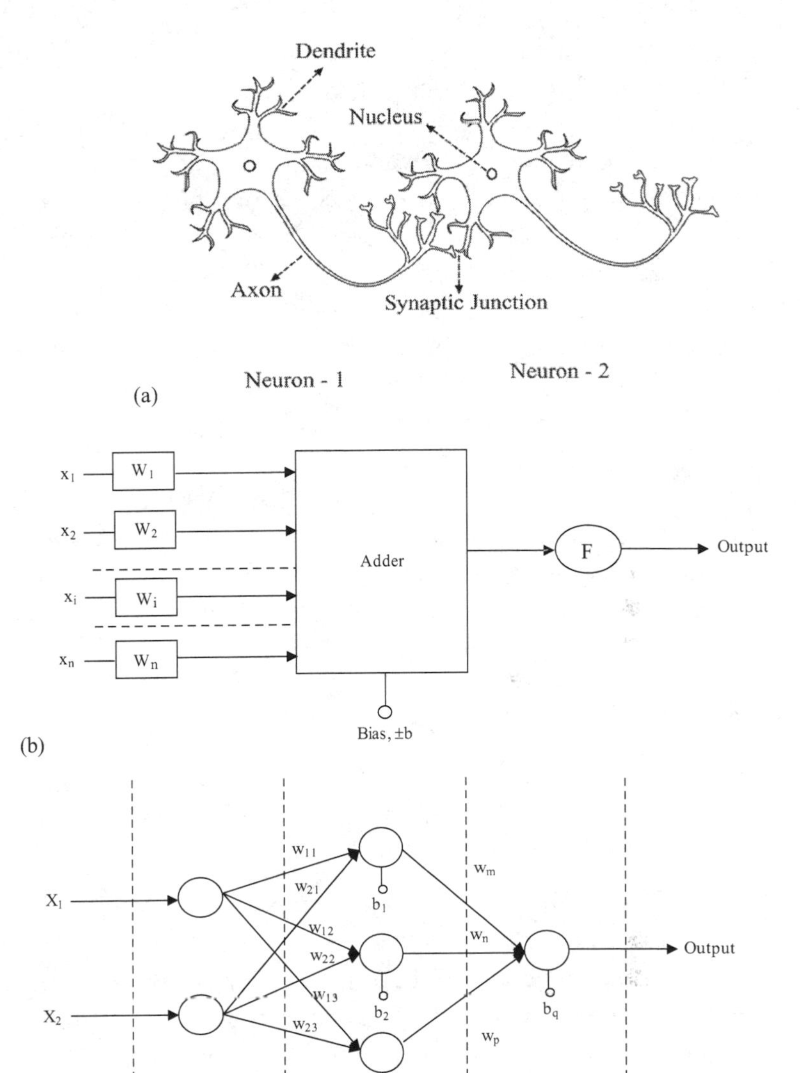

FIGURE 4.4 Representation of (a) biological neuron, (b) artificial neuron and (c) feed-forward neural network.

dendrites of the next neuron. This cumulative process is repeated, resulting in an all-round activity in the brain.

The simple structure and working of a neuron can be perceived as an adder of weighted information, as shown in Figure 4.4(b). The signal from the dendrite is changed or multiplied by the conductivity of the synaptic gap. The weighted information from all dendrites is aggregated at the neuron. The net contribution in a neuron is weighted against a bias voltage existing at the neuron. Further, this neuron sends out a signal, which is a function of the input to the neuron together with the bias voltage. From these descriptions, a representation of an artificial neuron network is given in Figure 4.4(b).

In Figure 4.4(b), $x_1, x_2, \ldots, x_i, \ldots, x_n$ are the input signals, each multiplied by individual weights, $w_1, w_2, \ldots, w_i, \ldots w_n$, and added together. The bias signal b is added or subtracted and the net value is mapped by using a proper function F.

Thus,

$$\text{Output} = F\left(x_1 w_1 + x_2 w_2 \cdots + x_i w_i + \cdots x_m w_m \pm b\right)$$

$$= F\left(\left(\sum_{i=l}^{n} x_i w_i\right) \pm b\right)$$

Depending on the application, there are a few functions F, such as linear, tansig and logsig.

A neural network consists of a finite number of interconnected artificial neurons. While numerous connections exist, one of the most commonly employed artificial neural networks is the feedforward network and such a scheme is depicted in Figure 4.4(c).

In Figure 4.4(c), each circle represents an artificial neuron and, as can be seen, the network is categorised into input, hidden and output layers. This is called a 2-3-1 configuration.

The number of neurons is problem dependent and the feedforward network in Figure 4.4(c) represents a two input-single output system. The neurons in the input layer do not contain any mapping function. Input values are multiplied by the weights w_{ij} where $i = 1$ to 2 and $j = 1$ to 3 and are passed on to the neurons in the hidden layer. The processed output from the hidden neurons is further multiplied by w_m, w_n, w_p and is transmitted to the output neuron. Each neuron is also provided with a suitable bias.

The term "training of neural network" is concerned with the determination of all weights and biases of the network, so that for a given input set, the output of the network equals the desired output. Thus, the task of training is to identify appropriate weights and biases to minimise the error between the actual output of the network and the desired output. Once the neural network is trained, it is capable of producing an output, which approximately equals the desired output, even for input data, which was not used during training. Thus, a well-trained network can approximate any non-linear functions with arbitrary accuracy. In recent years, neural networks have been successfully developed for many applications.

While neural networks possess several advantages in estimation and control application over conventional techniques, the major hindrance, at least in a few cases, is the lack of adequate input-output data required for training. With reduced information, it may not be possible to successfully train a neural network.

With a limited information base, however, it is possible to construct a fuzzy logic system. The information need not be exact. Existing knowledge can be readily expressed in linguistic form and a fuzzy logic system can be constructed. The fuzzy logic system can then be activated for numerous inputs and corresponding outputs can be determined. This simple input-output pattern can now be used to train a neural network. A neural network, trained in this way, is termed a "neuro-fuzzy system". Thus, the symbolic representation in fuzzy logic and the numerical processing of neural networks are integrated to form the powerful tool of neuro-fuzzy systems. In such a coupling, common characteristics of both schemes inherently exist. Neuro-fuzzy approaches have been successfully used in many applications, such as parameter estimation, process control and forecasting.

4.7 FUZZY GENETIC ALGORITHMS

In this section, the properties of evolutionary computing are further enhanced by fuzzification of a genetic algorithm (GA). A brief overview of a GA follows.

A GA is a general-purpose search algorithm driven by the basic principle of the biological evolution process. A GA is a powerful optimisation tool and is superior to conventional optimisation methods. A GA possesses several advantages, such as computational simplicity, derivative-free operation, reduced number of steps and a guaranteed near-optimal solution. This method has been proved to be very useful in domains that are not well understood or for solution spaces that are too large to be searched effectively by standard optimisation methods. The major difference between a GA and conventional optimisation methods is that a GA is initiated with a large number of possible solution sets, whereas conventional methods start with a single solution. Thus, to start the process of GA, a large number of solution sets are randomly generated. Each solution set is called a chromosome and each chromosome is a feasible solution to the problem. Since the chromosome embeds the solution set for an N-variable problem, each chromosome contains N different variables. Each variable in the chromosome is called a gene, and each gene is generally represented by binary digits ("1" and "0"). Each binary bit is named an allele. Thus, genes constitute chromosomes and the total number of chromosomes is termed the "population size". Generally, a population size of 20 to 30 is considered the optimal choice.

The solution set or chromosome is generally represented in binary form. For example, a value of 12 is represented as 1100 with a 4-bit span.

For a two-variable problem, let 7 and 3 represent the genes. Then, the chromosome containing 7 and 3 in binary form is 0111/0011.

Thus, for an N variable, if M is the number of bits for each variable (gene), there will be M * N bits for each chromosome.

The various steps of a GA are

- Selection
- Crossover
- Mutation

A flow chart of a simple GA is shown in Figure 4.5(a) and is explained as follows:

Step 1. Generation of solution set: An initial solution set (i.e. initial population) is randomly chosen considering the predetermined range of variables.

Step 2. Fitness evaluation: Each chromosome is evaluated to determine the "quality" or "goodness" of a solution. This is performed by defining a "fitness function". The definition of a fitness function is very important where the quality of the output is concerned and the fitness function varies from one problem to another. The fitness value of each chromosome is derived from the objective function of the optimisation task. Chromosomes that represent better solutions are awarded a higher fitness value to enable them to pass on to the next generation. Thus, a table-like format is developed with each chromosome and its associated fitness function value as a member.

Step 3. Selection: The process in which individual solutions (chromosomes) are selected to generate offspring is called parent selection or reproduction. The selection process models nature's survival of the fittest mechanism. Fitter solutions survive while weaker ones perish. There are different selection schemes, such as the proportionate selection scheme and the roulette wheel selection scheme. The basic idea is to collect chromosomes with a good fitness value so that they produce better offspring.

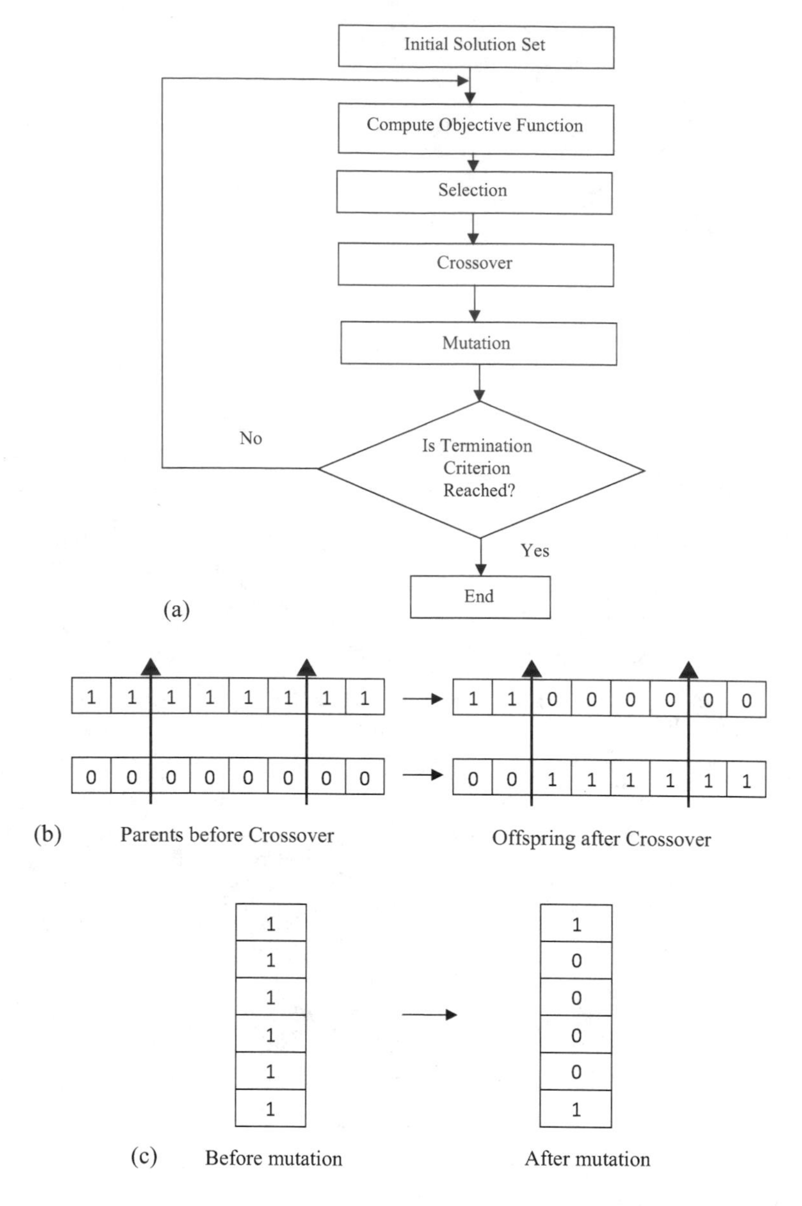

FIGURE 4.5 Steps of a GA (a) flow chart, (b) crossover and (c) mutation.

Step 4. Crossover: This operation comes immediately after selecting the parents. This is a major step in a GA, which is performed on a selected pair of chromosomes. Initially, two chromosomes from the reproduced population are considered and a crossover point is randomly selected. At the crossover point, the binary bits are interchanged between the selected pair resulting in two new chromosomes called offspring. This is illustrated in Figure 4.5(b). The vertically marked arrow line indicates a randomly chosen position for a crossover point. In a two-point crossover, two crossover points are randomly chosen and bits are swapped. This is shown in Figure 4.5(b).

It is important to mention that the crossover procedure is employed based on the value of the crossover possibility, P_c, which lies between 0 and 1. For this, a random number is generated in the range from 0 to 1 and if this number is greater than P_c, crossover is carried out. The probability of crossover P_c, which is also denoted the crossover rate, assumes significance in the diversification of a population. The determination of P_c is crucial and problem dependent.

Step 5. Mutation: Mutation is an important source of diversity in a GA and is performed after crossover. Mutation is applied to each bit of a chromosome and the mutation of a bit involves flipping it, that is changing a 0 to 1 and vice versa. The bits of a chromosome are independently mutated in the sense that the mutation of one bit does not affect the mutation of other bits. Just as P_c controls the crossover operation, another GA parameter, namely the probability of mutation P_m, gives the probability that a bit will be mutated; P_m is also called the mutation rate. The probability of mutation is generally low to prevent premature convergence of a GA. The judicial choice of P_m is very critical for the successful convergence of a GA to the global optimum. Figure 4.5(c) illustrates a mutation operation. Here, except the first and the last bits, all other bits are mutated.

4.8 FUZZY LOGIC FOR GENETIC ALGORITHMS

There are a number of ways in which GAs and fuzzy logic can be integrated. The most common approach is to use a GA to optimise the performance of a fuzzy system. An alternative approach is to use fuzzy logic techniques to improve the performance of the GA. A fuzzy genetic algorithm (FGA) is considered a GA that uses fuzzy-based techniques or fuzzy tools to improve the behaviour of the GA by modelling different GA components. In a fuzzy-controlled GA, the parameters of GA, namely crossover probability P_c and mutation probability P_m, are adjusted for an improved performance. An FGA employs a real-coded GA with multiple crossover and mutation operators. A schematic diagram is shown in Figure 4.6.

The FGA employs a combination of population statistics to control the adaptation of the probability of crossover and mutation and diversity statistics to control population diversity. At intervals or when the condition controlling population diversity has been

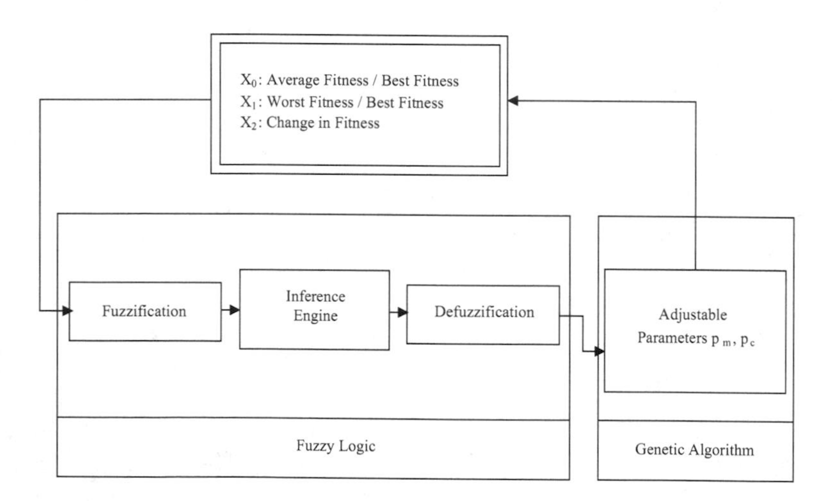

FIGURE 4.6 Fuzzy logic–tuned genetic algorithm.

reached, the GA passes the population details to the evaluation system, which calculates the necessary statistics. These are then passed to the fuzzy logic unit, where they are fuzzified, combined with other relative values and acted upon using the rules in the fuzzy knowledge rule base. Once complete, the integrated fuzzy sets are passed to be defuzzified, resulting in crisp values. These values are then returned to the GA, where they are used to provide alternative crossover operators and adjust the corresponding probabilities of crossover and mutations required. It is observed that a fuzzy-controlled GA is more efficient than a simple GA.

4.9 DC MOTOR SPEED CONTROL USING FUZZY LOGIC PRINCIPLE

A direct current (dc) motor is the best choice for many industrial applications due to its speed-torque characteristics. With the development of power semiconductor devices, a thyristorised armature voltage control system is the basic component of a motor speed control scheme. This is shown in Figure 1.2(a).

Here, the alternating current (ac) voltage supply is fed to a silicon controlled rectifier (SCR) converter, which converts the ac voltage to a dc voltage, with an average value of V_a. The amplitude of V_a depends on the SCR firing angle α and the motor speed varies with V_a. Thus, a variation in speed is achieved by varying α. With a variation in the load, the speed ω_r varies, causing a speed error, $\Delta\omega_r$, that is given by

$$\Delta\omega_r = \omega_r^* - \omega_r$$

This enables the controller to produce a variation in the SCR firing angle $\Delta\alpha$, leading to a variation in V_a and hence ω_r reaches ω_r^*.

The controller is typically a PI controller. However, the relationship between the converter output voltage, V_0, and the motor current, I_a, is non-linear during the discontinuous modes of the motor current. This makes the motor speed-torque characteristics non-linear, which is delineated in Figure 1.2(b).

Thus, a fixed gain controller will lead to speed fluctuations, which are highly undesirable. In order to achieve a satisfactory response of the drive system at all operating points, a fuzzy set theory can be utilised. A fuzzy controller can be developed using the following inputs and output:

Inputs

$$\text{Speed error } e = \omega_r^* - \omega_r$$

$$\text{Change of speed error } \Delta e = \Delta\omega_r(k) - \Delta\omega(k - I)$$

where k represents the instant at which the speed is sampled and (k – 1) refers to the previous sampling instant.

Output

The variation in the SCR firing angle $\Delta\alpha$.

Initially, a simulation of the whole closed loop system with a fixed gain PI controller designed at a typical operating point is carried out. The load torque is varied from no load to rated value and the universe of discourse of each input and output variable is determined. Then, the number of fuzzy sets is assumed at either 7, 9 or 11 and each is suitably labelled. Also, the domain of each fuzzy set is decided. Initially, an approximate rule base can be developed either on a trial-and-error basis or by using a transient response analysis. The initial rule base is used for simulation and is updated to achieve the best possible dynamic response. Once a robust rule base has been designed, the controller will act as an adaptive controller leading to an excellent dynamic response at all operating points.

4.10 FUZZY LOGIC–BASED WASHING MACHINE

Fuzzy logic principles are extensively employed in various consumer products and the sale of these products has increased in recent years. A typical application is the use of a fuzzy logic principle in automatic washing machines. While many manufacturers provide a variety of features, the underlying principle of a fuzzy logic–based washing machine is explained in this section.

The objective of washing clothes is to remove dirt particles. Thus, the input of the fuzzy control system is the quantum of dirt and its rate of change. The weight of clothes is another input. With these three inputs, the following are the control variables:

(1) Quantity of washing powder (this is generally displayed)

(2) Water quantity

(3) Water flow rate

(4) Washing time

(5) Rinsing time

(6) Spinning time

The weight of clothes is generally labelled LIGHT, MED and HEAVY. Similar names are employed for other variables too.

4.11 CONCLUSION

In this chapter, the application of fuzzy logic principles to various fields has been explained. Additionally, neural networks and genetic algorithms have been introduced and explained.

QUESTIONS

1. List and explain the general guidelines for the construction of fuzzy sets.
2. Why should adjacent fuzzy sets overlap?
3. What is the ideal value of the percentage overlap of two adjacent fuzzy sets?

4. Suggest labels for two fuzzy sets describing the speed of a motor.
5. How will you determine the number of fuzzy sets for a particular system variable?
6. Elucidate on the effect of the percentage of overlap between fuzzy sets.
7. If three fuzzy sets are used to describe atmospheric temperature as a variable, select appropriate names for each fuzzy set.
8. Three fuzzy sets are used to describe the following variables. Give suitable names for the fuzzy sets:
    (i) Height of students in a class
    (ii) Age of employees in a factory
    (iii) Speed of a motor
    (iv) Academic performance of students in a class
    (v) Temperature of an industrial process
    (vi) Traffic density in a city
    (vii) Running speed of a vehicle
9. Describe adaptive control and explain how this can be incorporated into fuzzy control systems.
10. Give two examples where adaptive control is essential.
11. Develop the mathematical formulation for fuzzy decision-making.
12. Illustrate fuzzy decision-making using the following example: student's choice of elective courses.

13. A research scholar working towards his PhD has to select ONE subject from the following four subjects for his coursework:

Subject	Department Offering the Course	Qualification of Teacher	Average Grade Obtained for the Past 5 years
X1: Basic power electronics	EEE	MTech	S
X2: Power system simulation	EEE	MTech	A
X3: DSP hardware and software	ECE	PhD	B
X4: Linear control theory	ICE	MTech	C

The research scholar must achieve at least a "C" grade and his broad area of research is "power electronic control of induction motors". Frame the problem as fuzzy decision-making and find the solution.

14. Mr. Pai wants to purchase a car. He generally employs a driver. He uses the car mostly to visit his home town which is around 300 km from his workplace. Mr. Pai's family includes his wife, two children and his elderly mother. Details of the cars available are as given. Formulate a fuzzy decision algorithm and select a suitable path.

Cars	Cost (x_1)	No. of Passengers (x_2)	Fuel Efficiency (km/ltr) (x_3)
c_1	1.1 lakhs	3	28
c_2	2.47 lakhs	4	22
c_3	3.79 lakhs	5	14
c_4	5.00 lakhs	4	12
c_5	6.63 lakhs	6	10

15. Sketch a biological neuron.

16. Build up artificial neural networks (ANNs) from the basic concept.

17. Give the structure of a 1-2-1 artificial neural network.
18. Elucidate on neuro-fuzzy systems.
19. What is the major advantage of neuro-fuzzy systems?
20. Which neuro-fuzzy system is preferred over fuzzy systems?
21. Give a detailed description of a genetic algorithm.
22. Define the term "population" in a genetic algorithm.
23. Define the fitness function in a genetic algorithm.
24. Describe the selection process in a genetic algorithm.
25. Explain the roulette wheel selection mechanism.
26. How is the survival of the fittest mechanism implemented in a genetic algorithm?
27. What is crossover in genetic algorithms?
28. What is mutation in a genetic algorithm?
29. Write the advantages of genetic algorithms over conventional optimisation methods.
30. What is crossover probability in a genetic algorithm and what is its range?
31. What is mutation probability and what is its usual range?
32. Give the methodology of a genetic algorithm and explain how fuzzy systems can be used to improve its performance.
33. Using a suitable optimisation problem, explain the different steps of a genetic algorithm to reach a solution.
34. Show the first two steps of a genetic algorithm–based approach towards the following problem:

$$\text{Max } y = \sin(x)$$

Subject to $0 \leq x \leq \pi$

35. Explain dc motor speed control using a fuzzy logic system.
36. Explain the fuzzy logic control in washing machines.
37. In an adaptive fuzzy control system, the variable u is described by fuzzy sets labelled ZERO, LOW, MED, HIGH and VERY HIGH, where the first and last are shouldered fuzzy sets and the remaining are triangular fuzzy sets. All fuzzy sets are symmetric and the universe of discourse is 0–800. When u = 300, perform fuzzification for the scaling factor unity and then 0.5.

Bibliography

K. Sundareswaran, High Performance AC Voltage Controller Fed Induction Motor Drive Using Fuzzy Logic Estimator. *Proceedings of the 2000 International Power Electronics Conference IPEC*, Tokyo, Japan, April 2000.

K. Sundareswaran, Application of Recent Control Techniques to the Design of Intelligent Controller for Energy Efficient Induction Motor Drive. PhD thesis, Bharathidasan University, 2001.

K. Sundareswaran and S. Palani, Fuzzy Logic Approach for Energy Efficient Voltage Controlled Induction Motor Drive. *Proceedings of 3rd IEEE International Conference on Power Electronics and Drive Systems, PEDS'99*, Vol. 1, pp. 552–554, Hong Kong, July 1999.